Grades
5-6

BUILDING

MATH SKILLS

I0510175

BY Brian Rhee

Over 2500 Practice Problems

Detailed Solutions

Legal Notice

Copyright © 2018 by Solomon Academy
Published by: Solomon Academy
First Edition
ISBN-13: 978-1717065827
ISBN-10: 1717065821

About This Book

This book is designed to help students build basic arithmetic and math skills. There are resources pages in the beginning of the book that illustrate how to do each type of problems. It is strongly recommended to go over the resources pages before solving practice problems.

This book contains 12 lessons with detailed solutions. Each lesson has ten practice worksheets which provides challenges to improve and strengthen students' math skills. Completing 12 lessons enables students to build their confidence and master their math skills.

About Author

Brian(Yeon) Rhee obtained a Masters of Arts Degree in Statistics at Columbia University, NY. He served as the Mathematical Statistician at the Bureau of Labor Statistics, DC. He is the Head Academic Director at Solomon Academy due to his devotion to the community coupled with his passion for teaching. His mission is to help students of all confidence level excel in academia to build a strong foundation in character, knowledge, and wisdom. Now, Solomon academy is known as the best academy specialized in Math in Northern Virginia.

Brian Rhee has published more than ten books. The titles of his books are 7 full-length practice tests for the AP Calculus AB/BC Multiple choice sections, AP Calculus, SAT 1 Math, SAT 2 Math level 2, 12 full-length practice tests for the SAT 2 Math Level 2, SHSAT/TJHSST Math workbook, and IAAT (Iowa Algebra Aptitude Test) Volume 1 and 2, CogAT form 7 Level 8, and NNAT 2 Level B Grade 1. He's currently working on other math books which will be introduced in the near future.

Brian Rhee has more than twenty years of teaching experience in math. He has been one of the most popular tutors among TJHSST (Thomas Jefferson High School For Science and Technology) students. Currently, he is developing many online math courses with www.masterprep.net for AP Calculus AB and BC, SAT 2 Math level 2 test, and other various math subjects.

SOLOMON ACADEMY

Solomon Academy is a prestigious institution of learning with numerous qualified teachers of various fields of education. Our mission is to thoroughly teach students of all ages and confidence levels, elevate skills to the highest standard of education, and provide them with all the tools and materials to succeed.

5723 Centre Square Drive
Centreville, VA 20120
Tel: 703-988-0019

Email: solomonacademyva@gmail.com
info@solomonacademy.net

CLASSES OFFERED

MATHEMATICS	TESTING	ENGLISH
1st-6th grade math	CogAt	1st-6th Reading
Algebra 1, 2	IAAT and SOL 7	1st-6th Writing
Geometry	TJHSST Prep	Essay Writing
Pre-Calculus	SAT/ACT Prep	SAT Writing
AP Calculus AB BC	SAT 2 Subject Tests	
AP Statistics	MathCounts	
Multivariate Calculus	AMC 10/12	

LEARN FROM THE AUTHOR

Private sessions with Brian Rhee is also available on the following subjects: SAT Math, SAT 2 Subject Math Level 2, Pre-Calculus, AP Calculus AB/BC, AP Statistics, IB SL/HL, Multivariate Calculus, Linear Algebra, AMC 8/10/12, and AIME.

Feel free to contact me at solomonacademyva@gmail.com

Acknowledgements

I wish to acknowledge my deepest appreciation to my wife, Sookyung, who has continuously given me wholehearted support, encouragement, and love. Without you, I could not have completed this book.

Thank you to my sons, Joshua and Jason, who have given me big smiles and inspiration. I love you all.

Thank you to Mr. Kwon from www.Masterprep.net, who has given me opportunities to develop online math courses for various math subjects.

Contents

Change the following fractions to percents.

Example: $\frac{1}{4} = 0.25 = 25\%$

1. $\frac{1}{4} = \underline{0.25} = \underline{25\%}$ 2. $\frac{1}{2} = \underline{0.5} = \underline{50\%}$

3. $\frac{1}{5} = \underline{\hspace{2cm}} = \underline{\hspace{2cm}}$ 4. $\frac{2}{5} = \underline{\hspace{2cm}} = \underline{\hspace{2cm}}$

5. $\frac{3}{4} = \underline{\hspace{2cm}} = \underline{\hspace{2cm}}$ 6. $\frac{4}{5} = \underline{\hspace{2cm}} = \underline{\hspace{2cm}}$

7. $\frac{3}{5} = \underline{\hspace{2cm}} = \underline{\hspace{2cm}}$ 8. $\frac{1}{10} = \underline{\hspace{2cm}} = \underline{\hspace{2cm}}$

9. $\frac{3}{10} = \underline{\hspace{2cm}} = \underline{\hspace{2cm}}$ 10. $\frac{2}{10} = \underline{\hspace{2cm}} = \underline{\hspace{2cm}}$

11. $\frac{1}{8} = \underline{\hspace{2cm}} = \underline{\hspace{2cm}}$ 12. $\frac{2}{8} = \underline{\hspace{2cm}} = \underline{\hspace{2cm}}$

13. $\frac{4}{8} = \underline{\hspace{2cm}} = \underline{\hspace{2cm}}$ 14. $\frac{3}{8} = \underline{\hspace{2cm}} = \underline{\hspace{2cm}}$

15. $\frac{5}{8} = \underline{\hspace{2cm}} = \underline{\hspace{2cm}}$ 16. $\frac{6}{8} = \underline{\hspace{2cm}} = \underline{\hspace{2cm}}$

Change the following fractions to percents.

Example: $\frac{1}{4} = 0.25 = 25\%$

1. $\frac{2}{5} = \underline{0.4} = \underline{40\%}$ 2. $\frac{3}{5} = \underline{0.6} = \underline{60\%}$

3. $\frac{1}{4} = \underline{\hspace{1.5cm}} = \underline{\hspace{1.5cm}}$ 4. $\frac{3}{4} = \underline{\hspace{1.5cm}} = \underline{\hspace{1.5cm}}$

5. $\frac{3}{2} = \underline{\hspace{1.5cm}} = \underline{\hspace{1.5cm}}$ 6. $\frac{5}{2} = \underline{\hspace{1.5cm}} = \underline{\hspace{1.5cm}}$

7. $\frac{2}{10} = \underline{\hspace{1.5cm}} = \underline{\hspace{1.5cm}}$ 8. $\frac{1}{8} = \underline{\hspace{1.5cm}} = \underline{\hspace{1.5cm}}$

9. $\frac{5}{8} = \underline{\hspace{1.5cm}} = \underline{\hspace{1.5cm}}$ 10. $\frac{3}{10} = \underline{\hspace{1.5cm}} = \underline{\hspace{1.5cm}}$

11. $\frac{6}{10} = \underline{\hspace{1.5cm}} = \underline{\hspace{1.5cm}}$ 12. $\frac{4}{10} = \underline{\hspace{1.5cm}} = \underline{\hspace{1.5cm}}$

13. $\frac{6}{8} = \underline{\hspace{1.5cm}} = \underline{\hspace{1.5cm}}$ 14. $\frac{7}{8} = \underline{\hspace{1.5cm}} = \underline{\hspace{1.5cm}}$

15. $\frac{6}{4} = \underline{\hspace{1.5cm}} = \underline{\hspace{1.5cm}}$ 16. $\frac{12}{10} = \underline{\hspace{1.5cm}} = \underline{\hspace{1.5cm}}$

Change the following fractions to percents.

Example: $\frac{1}{4} = 0.25 = 25\%$

1. $\frac{1}{8} = \underline{0.125} = \underline{12.5\%}$

2. $\frac{3}{5} = \underline{\hspace{1.5cm}} = \underline{\hspace{1.5cm}}$

3. $\frac{7}{10} = \underline{\hspace{1.5cm}} = \underline{\hspace{1.5cm}}$

4. $\frac{5}{4} = \underline{\hspace{1.5cm}} = \underline{\hspace{1.5cm}}$

5. $\frac{5}{2} = \underline{\hspace{1.5cm}} = \underline{\hspace{1.5cm}}$

6. $\frac{9}{10} = \underline{\hspace{1.5cm}} = \underline{\hspace{1.5cm}}$

7. $\frac{3}{8} = \underline{\hspace{1.5cm}} = \underline{\hspace{1.5cm}}$

8. $\frac{5}{8} = \underline{\hspace{1.5cm}} = \underline{\hspace{1.5cm}}$

9. $\frac{3}{4} = \underline{\hspace{1.5cm}} = \underline{\hspace{1.5cm}}$

10. $\frac{3}{2} = \underline{\hspace{1.5cm}} = \underline{\hspace{1.5cm}}$

11. $\frac{4}{5} = \underline{\hspace{1.5cm}} = \underline{\hspace{1.5cm}}$

12. $\frac{7}{8} = \underline{\hspace{1.5cm}} = \underline{\hspace{1.5cm}}$

13. $\frac{3}{20} = \underline{\hspace{1.5cm}} = \underline{\hspace{1.5cm}}$

14. $\frac{1}{20} = \underline{\hspace{1.5cm}} = \underline{\hspace{1.5cm}}$

15. $\frac{13}{10} = \underline{\hspace{1.5cm}} = \underline{\hspace{1.5cm}}$

16. $\frac{7}{4} = \underline{\hspace{1.5cm}} = \underline{\hspace{1.5cm}}$

Change the following fractions to percents.

Example: $\frac{1}{4} = 0.25 = 25\%$

1. $\frac{9}{12} =$ _____ = _____

2. $\frac{6}{10} =$ _____ = _____

3. $\frac{4}{10} =$ _____ = _____

4. $\frac{6}{5} =$ _____ = _____

5. $\frac{7}{2} =$ _____ = _____

6. $\frac{7}{4} =$ _____ = _____

7. $\frac{7}{5} =$ _____ = _____

8. $\frac{3}{20} =$ _____ = _____

9. $\frac{12}{8} =$ _____ = _____

10. $\frac{7}{8} =$ _____ = _____

11. $\frac{14}{10} =$ _____ = _____

12. $\frac{7}{20} =$ _____ = _____

13. $\frac{5}{4} =$ _____ = _____

14. $\frac{12}{20} =$ _____ = _____

15. $\frac{9}{20} =$ _____ = _____

16. $\frac{11}{20} =$ _____ = _____

Change the following fractions to percents.

Example: $\frac{1}{4} = 0.25 = 25\%$

1. $\frac{1}{5} = $ _____ $= $ _____

2. $\frac{4}{8} = $ _____ $= $ _____

3. $\frac{9}{10} = $ _____ $= $ _____

4. $\frac{7}{4} = $ _____ $= $ _____

5. $\frac{9}{6} = $ _____ $= $ _____

6. $\frac{8}{5} = $ _____ $= $ _____

7. $\frac{9}{15} = $ _____ $= $ _____

8. $\frac{7}{5} = $ _____ $= $ _____

9. $\frac{4}{5} = $ _____ $= $ _____

10. $\frac{7}{10} = $ _____ $= $ _____

11. $\frac{9}{20} = $ _____ $= $ _____

12. $\frac{7}{20} = $ _____ $= $ _____

13. $\frac{13}{10} = $ _____ $= $ _____

14. $\frac{11}{10} = $ _____ $= $ _____

15. $\frac{12}{8} = $ _____ $= $ _____

16. $\frac{13}{20} = $ _____ $= $ _____

Name:

Change the following fractions to percents.

Example: $\frac{1}{4} = 0.25 = 25\%$

1. $\dfrac{14}{10} =$ _____ = _____

2. $\dfrac{8}{10} =$ _____ = _____

3. $\dfrac{6}{2} =$ _____ = _____

4. $\dfrac{5}{2} =$ _____ = _____

5. $\dfrac{7}{10} =$ _____ = _____

6. $\dfrac{6}{10} =$ _____ = _____

7. $\dfrac{6}{20} =$ _____ = _____

8. $\dfrac{3}{25} =$ _____ = _____

9. $\dfrac{9}{20} =$ _____ = _____

10. $\dfrac{16}{10} =$ _____ = _____

11. $\dfrac{8}{5} =$ _____ = _____

12. $\dfrac{7}{20} =$ _____ = _____

13. $\dfrac{15}{6} =$ _____ = _____

14. $\dfrac{9}{5} =$ _____ = _____

15. $\dfrac{19}{20} =$ _____ = _____

16. $\dfrac{17}{20} =$ _____ = _____

Change the following fractions to percents.

Example: $\frac{1}{4} = 0.25 = 25\%$

1. $\dfrac{7}{8} =$ _____ $=$ _____

2. $\dfrac{3}{8} =$ _____ $=$ _____

3. $\dfrac{6}{10} =$ _____ $=$ _____

4. $\dfrac{9}{10} =$ _____ $=$ _____

5. $\dfrac{16}{25} =$ _____ $=$ _____

6. $\dfrac{12}{10} =$ _____ $=$ _____

7. $\dfrac{9}{8} =$ _____ $=$ _____

8. $\dfrac{11}{8} =$ _____ $=$ _____

9. $\dfrac{7}{5} =$ _____ $=$ _____

10. $\dfrac{12}{5} =$ _____ $=$ _____

11. $\dfrac{13}{20} =$ _____ $=$ _____

12. $\dfrac{7}{25} =$ _____ $=$ _____

13. $\dfrac{20}{8} =$ _____ $=$ _____

14. $\dfrac{17}{20} =$ _____ $=$ _____

15. $\dfrac{40}{20} =$ _____ $=$ _____

16. $\dfrac{22}{20} =$ _____ $=$ _____

Change the following fractions to percents.

Example:　　$\frac{1}{4} = 0.25 = 25\%$

1. $\frac{8}{10} =$ _____ = _____　　　　　2. $\frac{6}{4} =$ _____ = _____

3. $\frac{14}{10} =$ _____ = _____　　　　　4. $\frac{8}{2} =$ _____ = _____

5. $\frac{9}{20} =$ _____ = _____　　　　　6. $\frac{10}{8} =$ _____ = _____

7. $\frac{13}{8} =$ _____ = _____　　　　　8. $\frac{17}{20} =$ _____ = _____

9. $\frac{24}{20} =$ _____ = _____　　　　10. $\frac{30}{20} =$ _____ = _____

11. $\frac{13}{5} =$ _____ = _____　　　　12. $\frac{16}{5} =$ _____ = _____

13. $\frac{3}{20} =$ _____ = _____　　　　14. $\frac{15}{8} =$ _____ = _____

15. $\frac{9}{25} =$ _____ = _____　　　　16. $\frac{11}{25} =$ _____ = _____

Change the following fractions to percents.

Example: $\frac{1}{4} = 0.25 = 25\%$

1. $\frac{9}{4} = $ _____ $ = $ _____

2. $\frac{3}{8} = $ _____ $ = $ _____

3. $\frac{6}{10} = $ _____ $ = $ _____

4. $\frac{4}{5} = $ _____ $ = $ _____

5. $\frac{25}{10} = $ _____ $ = $ _____

6. $\frac{5}{4} = $ _____ $ = $ _____

7. $\frac{20}{8} = $ _____ $ = $ _____

8. $\frac{11}{8} = $ _____ $ = $ _____

9. $\frac{7}{20} = $ _____ $ = $ _____

10. $\frac{15}{20} = $ _____ $ = $ _____

11. $\frac{4}{25} = $ _____ $ = $ _____

12. $\frac{35}{10} = $ _____ $ = $ _____

13. $\frac{17}{10} = $ _____ $ = $ _____

14. $\frac{36}{20} = $ _____ $ = $ _____

15. $\frac{40}{20} = $ _____ $ = $ _____

16. $\frac{17}{5} = $ _____ $ = $ _____

Change the following fractions to percents.

Example: $\frac{1}{4} = 0.25 = 25\%$

1. $\frac{11}{10} = $ _____ = _____

2. $\frac{6}{20} = $ _____ = _____

3. $\frac{8}{2} = $ _____ = _____

4. $\frac{9}{2} = $ _____ = _____

5. $\frac{9}{8} = $ _____ = _____

6. $\frac{9}{6} = $ _____ = _____

7. $\frac{12}{8} = $ _____ = _____

8. $\frac{25}{10} = $ _____ = _____

9. $\frac{7}{25} = $ _____ = _____

10. $\frac{14}{5} = $ _____ = _____

11. $\frac{38}{20} = $ _____ = _____

12. $\frac{39}{20} = $ _____ = _____

13. $\frac{18}{5} = $ _____ = _____

14. $\frac{17}{20} = $ _____ = _____

15. $\frac{11}{20} = $ _____ = _____

16. $\frac{18}{6} = $ _____ = _____

Change the following percents to fractions.

Example: $75\% = \frac{75}{100} = \frac{3}{4}$

1. $25\% = \dfrac{25}{100} = \dfrac{1}{4}$

2. $20\% = \dfrac{20}{100} = \dfrac{1}{5}$

3. $50\% = \underline{\qquad} = \underline{\qquad}$

4. $30\% = \underline{\qquad} = \underline{\qquad}$

5. $35\% = \underline{\qquad} = \underline{\qquad}$

6. $45\% = \underline{\qquad} = \underline{\qquad}$

7. $60\% = \underline{\qquad} = \underline{\qquad}$

8. $55\% = \underline{\qquad} = \underline{\qquad}$

9. $40\% = \underline{\qquad} = \underline{\qquad}$

10. $70\% = \underline{\qquad} = \underline{\qquad}$

11. $4\% = \underline{\qquad} = \underline{\qquad}$

12. $12\% = \underline{\qquad} = \underline{\qquad}$

13. $32\% = \underline{\qquad} = \underline{\qquad}$

14. $80\% = \underline{\qquad} = \underline{\qquad}$

15. $90\% = \underline{\qquad} = \underline{\qquad}$

16. $96\% = \underline{\qquad} = \underline{\qquad}$

Change the following percents to fractions.

Example: $75\% = \frac{75}{100} = \frac{3}{4}$

1. $50\% =$ _____ = _____

2. $85\% =$ _____ = _____

3. $35\% =$ _____ = _____

4. $15\% =$ _____ = _____

5. $80\% =$ _____ = _____

6. $75\% =$ _____ = _____

7. $45\% =$ _____ = _____

8. $90\% =$ _____ = _____

9. $44\% =$ _____ = _____

10. $55\% =$ _____ = _____

11. $95\% =$ _____ = _____

12. $56\% =$ _____ = _____

13. $72\% =$ _____ = _____

14. $48\% =$ _____ = _____

15. $65\% =$ _____ = _____

16. $96\% =$ _____ = _____

Change the following percents to fractions.

Example: $75\% = \frac{75}{100} = \frac{3}{4}$

1. $25\% = $ _____ $= $ _____   2. $50\% = $ _____ $= $ _____

3. $16\% = $ _____ $= $ _____   4. $70\% = $ _____ $= $ _____

5. $45\% = $ _____ $= $ _____   6. $65\% = $ _____ $= $ _____

7. $42\% = $ _____ $= $ _____   8. $35\% = $ _____ $= $ _____

9. $90\% = $ _____ $= $ _____   10. $88\% = $ _____ $= $ _____

11. $20\% = $ _____ $= $ _____   12. $64\% = $ _____ $= $ _____

13. $92\% = $ _____ $= $ _____   14. $8\% = $ _____ $= $ _____

15. $95\% = $ _____ $= $ _____   16. $15\% = $ _____ $= $ _____

Change the following percents to fractions.

Example: $75\% = \frac{75}{100} = \frac{3}{4}$

1. $82\% = $ _____ $ = $ _____

2. $35\% = $ _____ $ = $ _____

3. $125\% = $ _____ $ = $ _____

4. $150\% = $ _____ $ = $ _____

5. $15\% = $ _____ $ = $ _____

6. $45\% = $ _____ $ = $ _____

7. $250\% = $ _____ $ = $ _____

8. $64\% = $ _____ $ = $ _____

9. $72\% = $ _____ $ = $ _____

10. $175\% = $ _____ $ = $ _____

11. $55\% = $ _____ $ = $ _____

12. $65\% = $ _____ $ = $ _____

13. $170\% = $ _____ $ = $ _____

14. $275\% = $ _____ $ = $ _____

15. $24\% = $ _____ $ = $ _____

16. $34\% = $ _____ $ = $ _____

Change the following percents to fractions.

Example: $75\% = \frac{75}{100} = \frac{3}{4}$

1. $175\% = $ _____ = _____   2. $120\% = $ _____ = _____

3. $48\% = $ _____ = _____   4. $65\% = $ _____ = _____

5. $110\% = $ _____ = _____   6. $5\% = $ _____ = _____

7. $14\% = $ _____ = _____   8. $8\% = $ _____ = _____

9. $92\% = $ _____ = _____   10. $180\% = $ _____ = _____

11. $190\% = $ _____ = _____   12. $325\% = $ _____ = _____

13. $64\% = $ _____ = _____   14. $210\% = $ _____ = _____

15. $24\% = $ _____ = _____   16. $76\% = $ _____ = _____

Change the following percents to fractions.

Example: $75\% = \frac{75}{100} = \frac{3}{4}$

1. $98\% =$ _____ $=$ _____

2. $130\% =$ _____ $=$ _____

3. $88\% =$ _____ $=$ _____

4. $37.5\% =$ _____ $=$ _____

5. $225\% =$ _____ $=$ _____

6. $150\% =$ _____ $=$ _____

7. $12.5\% =$ _____ $=$ _____

8. $44\% =$ _____ $=$ _____

9. $175\% =$ _____ $=$ _____

10. $62.5\% =$ _____ $=$ _____

11. $155\% =$ _____ $=$ _____

12. $125\% =$ _____ $=$ _____

13. $87.5\% =$ _____ $=$ _____

14. $165\% =$ _____ $=$ _____

15. $105\% =$ _____ $=$ _____

16. $220\% =$ _____ $=$ _____

Change the following percents to fractions.

Example: $75\% = \frac{75}{100} = \frac{3}{4}$

1. $62.5\% =$ _____ $=$ _____

2. $65\% =$ _____ $=$ _____

3. $170\% =$ _____ $=$ _____

4. $140\% =$ _____ $=$ _____

5. $200\% =$ _____ $=$ _____

6. $87.5\% =$ _____ $=$ _____

7. $145\% =$ _____ $=$ _____

8. $5\% =$ _____ $=$ _____

9. $95\% =$ _____ $=$ _____

10. $225\% =$ _____ $=$ _____

11. $144\% =$ _____ $=$ _____

12. $155\% =$ _____ $=$ _____

13. $12.5\% =$ _____ $=$ _____

14. $124\% =$ _____ $=$ _____

15. $54\% =$ _____ $=$ _____

16. $195\% =$ _____ $=$ _____

Change the following percents to fractions.

Example: $75\% = \frac{75}{100} = \frac{3}{4}$

1. $28\% = $ _____ $ = $ _____
2. $52\% = $ _____ $ = $ _____

3. $35\% = $ _____ $ = $ _____
4. $36\% = $ _____ $ = $ _____

5. $190\% = $ _____ $ = $ _____
6. $85\% = $ _____ $ = $ _____

7. $250\% = $ _____ $ = $ _____
8. $128\% = $ _____ $ = $ _____

9. $175\% = $ _____ $ = $ _____
10. $300\% = $ _____ $ = $ _____

11. $115\% = $ _____ $ = $ _____
12. $225\% = $ _____ $ = $ _____

13. $108\% = $ _____ $ = $ _____
14. $188\% = $ _____ $ = $ _____

15. $280\% = $ _____ $ = $ _____
16. $152\% = $ _____ $ = $ _____

Change the following percents to fractions.

Example: $75\% = \frac{75}{100} = \frac{3}{4}$

1. $56\% = $ _____ $ = $ _____

2. $250\% = $ _____ $ = $ _____

3. $130\% = $ _____ $ = $ _____

4. $64\% = $ _____ $ = $ _____

5. $325\% = $ _____ $ = $ _____

6. $45\% = $ _____ $ = $ _____

7. $112\% = $ _____ $ = $ _____

8. $96\% = $ _____ $ = $ _____

9. $35\% = $ _____ $ = $ _____

10. $102\% = $ _____ $ = $ _____

11. $165\% = $ _____ $ = $ _____

12. $375\% = $ _____ $ = $ _____

13. $148\% = $ _____ $ = $ _____

14. $16\% = $ _____ $ = $ _____

15. $62.5\% = $ _____ $ = $ _____

16. $152\% = $ _____ $ = $ _____

Change the following percents to fractions.

Example: $75\% = \frac{75}{100} = \frac{3}{4}$

1. $275\% = $ _____ $= $ _____

2. $375\% = $ _____ $= $ _____

3. $146\% = $ _____ $= $ _____

4. $220\% = $ _____ $= $ _____

5. $115\% = $ _____ $= $ _____

6. $96\% = $ _____ $= $ _____

7. $196\% = $ _____ $= $ _____

8. $164\% = $ _____ $= $ _____

9. $400\% = $ _____ $= $ _____

10. $138\% = $ _____ $= $ _____

11. $150\% = $ _____ $= $ _____

12. $105\% = $ _____ $= $ _____

13. $215\% = $ _____ $= $ _____

14. $172\% = $ _____ $= $ _____

15. $185\% = $ _____ $= $ _____

16. $350\% = $ _____ $= $ _____

Change the following mixed numbers to improper fractions.

Example: $2\frac{1}{2} = 2 + \frac{1}{2} = \frac{4}{2} + \frac{1}{2} = \frac{5}{2}$

1. $1\frac{2}{3} =$ 2. $1\frac{1}{4} =$ 3. $2\frac{1}{5} =$

4. $2\frac{2}{5} =$ 5. $1\frac{1}{3} =$ 6. $2\frac{2}{3} =$

7. $1\frac{2}{7} =$ 8. $1\frac{3}{7} =$ 9. $2\frac{3}{7} =$

10. $1\frac{3}{4} =$ 11. $2\frac{3}{5} =$ 12. $1\frac{3}{5} =$

13. $2\frac{4}{5} =$ 14. $2\frac{2}{7} =$ 15. $1\frac{6}{7} =$

16. $1\frac{1}{6} =$ 17. $1\frac{4}{7} =$ 18. $2\frac{1}{7} =$

19. $2\frac{5}{7} =$ 20. $1\frac{5}{6} =$ 21. $3\frac{5}{8} =$

22. $1\frac{7}{8} =$ 23. $2\frac{5}{8} =$ 24. $3\frac{5}{6} =$

Change the following mixed numbers to improper fractions.

Example: $2\frac{1}{2} = 2 + \frac{1}{2} = \frac{4}{2} + \frac{1}{2} = \frac{5}{2}$

1. $1\frac{2}{3} =$ 2. $1\frac{1}{6} =$ 3. $2\frac{4}{5} =$

4. $2\frac{1}{4} =$ 5. $1\frac{1}{3} =$ 6. $3\frac{1}{2} =$

7. $1\frac{2}{5} =$ 8. $1\frac{3}{4} =$ 9. $2\frac{4}{9} =$

10. $1\frac{3}{7} =$ 11. $2\frac{3}{5} =$ 12. $1\frac{6}{7} =$

13. $2\frac{5}{6} =$ 14. $2\frac{4}{7} =$ 15. $1\frac{5}{9} =$

16. $3\frac{1}{8} =$ 17. $3\frac{3}{8} =$ 18. $2\frac{7}{10} =$

19. $2\frac{1}{9} =$ 20. $1\frac{2}{9} =$ 21. $2\frac{5}{8} =$

22. $3\frac{3}{10} =$ 23. $3\frac{7}{8} =$ 24. $3\frac{5}{7} =$

Change the following mixed numbers to improper fractions.

Example: $2\dfrac{1}{2} = 2 + \dfrac{1}{2} = \dfrac{4}{2} + \dfrac{1}{2} = \dfrac{5}{2}$

1. $3\dfrac{1}{6} =$

2. $2\dfrac{5}{6} =$

3. $1\dfrac{6}{7} =$

4. $3\dfrac{3}{4} =$

5. $3\dfrac{1}{4} =$

6. $3\dfrac{3}{10} =$

7. $1\dfrac{1}{10} =$

8. $2\dfrac{5}{6} =$

9. $2\dfrac{4}{9} =$

10. $2\dfrac{3}{4} =$

11. $2\dfrac{5}{8} =$

12. $1\dfrac{8}{9} =$

13. $2\dfrac{3}{8} =$

14. $3\dfrac{7}{10} =$

15. $2\dfrac{6}{7} =$

16. $2\dfrac{1}{8} =$

17. $1\dfrac{5}{7} =$

18. $3\dfrac{1}{8} =$

19. $3\dfrac{1}{9} =$

20. $2\dfrac{7}{8} =$

21. $3\dfrac{2}{7} =$

22. $1\dfrac{9}{10} =$

23. $2\dfrac{7}{10} =$

24. $3\dfrac{1}{10} =$

31

Lesson 3-4   Changing mixed numbers to improper fractions

Change the following mixed numbers to improper fractions.

Example: $2\frac{1}{2} = 2 + \frac{1}{2} = \frac{4}{2} + \frac{1}{2} = \frac{5}{2}$

1. $2\frac{1}{5} =$ 2. $1\frac{5}{6} =$ 3. $3\frac{4}{5} =$

4. $4\frac{3}{4} =$ 5. $3\frac{1}{4} =$ 6. $2\frac{3}{7} =$

7. $3\frac{1}{6} =$ 8. $2\frac{2}{5} =$ 9. $3\frac{5}{8} =$

10. $3\frac{2}{7} =$ 11. $2\frac{1}{7} =$ 12. $4\frac{4}{7} =$

13. $1\frac{3}{8} =$ 14. $4\frac{3}{8} =$ 15. $2\frac{5}{9} =$

16. $4\frac{2}{9} =$ 17. $3\frac{4}{9} =$ 18. $3\frac{6}{7} =$

19. $3\frac{3}{11} =$ 20. $4\frac{3}{10} =$ 21. $2\frac{7}{11} =$

22. $3\frac{7}{10} =$ 23. $4\frac{5}{7} =$ 24. $4\frac{7}{8} =$

Change the following mixed numbers to improper fractions.

Example: $2\frac{1}{2} = 2 + \frac{1}{2} = \frac{4}{2} + \frac{1}{2} = \frac{5}{2}$

1. $2\frac{1}{4} =$

2. $3\frac{1}{6} =$

3. $2\frac{6}{7} =$

4. $4\frac{5}{6} =$

5. $2\frac{5}{7} =$

6. $3\frac{5}{9} =$

7. $3\frac{3}{5} =$

8. $4\frac{2}{5} =$

9. $5\frac{4}{5} =$

10. $5\frac{2}{7} =$

11. $4\frac{3}{4} =$

12. $4\frac{7}{9} =$

13. $2\frac{4}{9} =$

14. $5\frac{2}{9} =$

15. $5\frac{9}{10} =$

16. $3\frac{8}{9} =$

17. $2\frac{3}{10} =$

18. $3\frac{3}{11} =$

19. $5\frac{7}{10} =$

20. $3\frac{2}{11} =$

21. $4\frac{5}{12} =$

22. $4\frac{4}{11} =$

23. $5\frac{1}{12} =$

24. $5\frac{1}{13} =$

Change the following mixed numbers to improper fractions.

Example: $2\frac{1}{2} = 2 + \frac{1}{2} = \frac{4}{2} + \frac{1}{2} = \frac{5}{2}$

1. $2\frac{2}{13} =$ 2. $3\frac{5}{8} =$ 3. $3\frac{4}{9} =$

4. $4\frac{7}{8} =$ 5. $2\frac{2}{5} =$ 6. $4\frac{5}{6} =$

7. $5\frac{1}{5} =$ 8. $3\frac{3}{13} =$ 9. $5\frac{4}{5} =$

10. $6\frac{1}{4} =$ 11. $5\frac{2}{9} =$ 12. $4\frac{3}{8} =$

13. $4\frac{3}{10} =$ 14. $5\frac{7}{10} =$ 15. $6\frac{3}{4} =$

16. $5\frac{5}{9} =$ 17. $4\frac{7}{9} =$ 18. $5\frac{7}{13} =$

19. $3\frac{10}{11} =$ 20. $5\frac{9}{11} =$ 21. $6\frac{9}{10} =$

22. $4\frac{1}{14} =$ 23. $3\frac{3}{14} =$ 24. $4\frac{8}{11} =$

Change the following mixed numbers to improper fractions.

Example: $2\dfrac{1}{2} = 2 + \dfrac{1}{2} = \dfrac{4}{2} + \dfrac{1}{2} = \dfrac{5}{2}$

1. $3\dfrac{3}{7} =$

2. $4\dfrac{2}{3} =$

3. $4\dfrac{3}{4} =$

4. $5\dfrac{5}{6} =$

5. $6\dfrac{7}{10} =$

6. $6\dfrac{2}{5} =$

7. $4\dfrac{3}{10} =$

8. $5\dfrac{6}{7} =$

9. $3\dfrac{4}{7} =$

10. $5\dfrac{3}{5} =$

11. $5\dfrac{4}{5} =$

12. $5\dfrac{4}{11} =$

13. $6\dfrac{2}{9} =$

14. $4\dfrac{5}{9} =$

15. $4\dfrac{11}{12} =$

16. $4\dfrac{5}{12} =$

17. $6\dfrac{2}{11} =$

18. $6\dfrac{5}{7} =$

19. $4\dfrac{7}{11} =$

20. $3\dfrac{9}{13} =$

21. $6\dfrac{4}{9} =$

22. $6\dfrac{4}{13} =$

23. $4\dfrac{7}{12} =$

24. $5\dfrac{8}{13} =$

Change the following mixed numbers to improper fractions.

Example: $2\frac{1}{2} = 2 + \frac{1}{2} = \frac{4}{2} + \frac{1}{2} = \frac{5}{2}$

1. $6\frac{3}{8} =$ 2. $6\frac{3}{5} =$ 3. $6\frac{1}{6} =$

4. $5\frac{1}{6} =$ 5. $4\frac{2}{11} =$ 6. $5\frac{3}{11} =$

7. $7\frac{2}{5} =$ 8. $5\frac{5}{8} =$ 9. $6\frac{4}{5} =$

10. $6\frac{2}{7} =$ 11. $4\frac{4}{7} =$ 12. $4\frac{5}{14} =$

13. $5\frac{9}{14} =$ 14. $7\frac{3}{4} =$ 15. $7\frac{7}{8} =$

16. $8\frac{2}{3} =$ 17. $5\frac{5}{6} =$ 18. $6\frac{7}{11} =$

19. $5\frac{11}{12} =$ 20. $4\frac{2}{13} =$ 21. $5\frac{3}{7} =$

22. $7\frac{8}{15} =$ 23. $7\frac{7}{15} =$ 24. $7\frac{3}{14} =$

Change the following mixed numbers to improper fractions.

Example: $2\frac{1}{2} = 2 + \frac{1}{2} = \frac{4}{2} + \frac{1}{2} = \frac{5}{2}$

1. $5\frac{1}{7} =$

2. $5\frac{8}{11} =$

3. $8\frac{1}{6} =$

4. $6\frac{2}{5} =$

5. $6\frac{2}{3} =$

6. $6\frac{7}{12} =$

7. $6\frac{4}{11} =$

8. $5\frac{5}{7} =$

9. $5\frac{4}{13} =$

10. $7\frac{5}{6} =$

11. $7\frac{8}{9} =$

12. $6\frac{11}{15} =$

13. $5\frac{7}{16} =$

14. $8\frac{3}{5} =$

15. $8\frac{3}{4} =$

16. $8\frac{3}{7} =$

17. $6\frac{6}{15} =$

18. $6\frac{7}{13} =$

19. $9\frac{7}{9} =$

20. $7\frac{5}{14} =$

21. $7\frac{6}{7} =$

22. $8\frac{8}{17} =$

23. $9\frac{9}{16} =$

24. $9\frac{4}{5} =$

Change the following mixed numbers to improper fractions.

Example: $2\frac{1}{2} = 2 + \frac{1}{2} = \frac{4}{2} + \frac{1}{2} = \frac{5}{2}$

1. $8\frac{3}{4} =$

2. $8\frac{2}{7} =$

3. $8\frac{7}{10} =$

4. $7\frac{5}{6} =$

5. $10\frac{2}{3} =$

6. $9\frac{3}{5} =$

7. $9\frac{4}{5} =$

8. $7\frac{9}{20} =$

9. $7\frac{5}{12} =$

10. $10\frac{5}{8} =$

11. $9\frac{8}{11} =$

12. $9\frac{11}{20} =$

13. $8\frac{13}{18} =$

14. $9\frac{6}{7} =$

15. $10\frac{2}{5} =$

16. $8\frac{9}{16} =$

17. $7\frac{4}{19} =$

18. $9\frac{8}{17} =$

19. $9\frac{9}{19} =$

20. $8\frac{8}{15} =$

21. $10\frac{4}{7} =$

22. $10\frac{17}{20} =$

23. $10\frac{7}{8} =$

24. $8\frac{9}{13} =$

Change the following improper fractions to mixed numbers.

Example: $\dfrac{13}{5} = \dfrac{10}{5} + \dfrac{3}{5} = 2 + \dfrac{3}{5} = 2\dfrac{3}{5}$

1. $\dfrac{11}{3} =$ 2. $\dfrac{9}{5} =$ 3. $\dfrac{5}{3} =$

4. $\dfrac{7}{4} =$ 5. $\dfrac{10}{3} =$ 6. $\dfrac{11}{5} =$

7. $\dfrac{14}{5} =$ 8. $\dfrac{13}{6} =$ 9. $\dfrac{13}{3} =$

10. $\dfrac{11}{8} =$ 11. $\dfrac{13}{4} =$ 12. $\dfrac{8}{6} =$

13. $\dfrac{11}{9} =$ 14. $\dfrac{9}{7} =$ 15. $\dfrac{15}{8} =$

16. $\dfrac{9}{6} =$ 17. $\dfrac{10}{8} =$ 18. $\dfrac{11}{7} =$

19. $\dfrac{13}{10} =$ 20. $\dfrac{13}{9} =$ 21. $\dfrac{19}{9} =$

22. $\dfrac{14}{11} =$ 23. $\dfrac{16}{12} =$ 24. $\dfrac{15}{12} =$

Change the following improper fractions to mixed numbers.

Example: $\dfrac{13}{5} = \dfrac{10}{5} + \dfrac{3}{5} = 2 + \dfrac{3}{5} = 2\dfrac{3}{5}$

1. $\dfrac{17}{5} =$

2. $\dfrac{10}{7} =$

3. $\dfrac{7}{6} =$

4. $\dfrac{6}{4} =$

5. $\dfrac{15}{2} =$

6. $\dfrac{19}{5} =$

7. $\dfrac{13}{7} =$

8. $\dfrac{11}{4} =$

9. $\dfrac{13}{3} =$

10. $\dfrac{19}{8} =$

11. $\dfrac{29}{5} =$

12. $\dfrac{29}{6} =$

13. $\dfrac{20}{9} =$

14. $\dfrac{21}{8} =$

15. $\dfrac{15}{7} =$

16. $\dfrac{39}{10} =$

17. $\dfrac{33}{10} =$

18. $\dfrac{44}{9} =$

19. $\dfrac{27}{12} =$

20. $\dfrac{31}{9} =$

21. $\dfrac{26}{11} =$

22. $\dfrac{20}{11} =$

23. $\dfrac{38}{11} =$

24. $\dfrac{47}{10} =$

Change the following improper fractions to mixed numbers.

Example: $\dfrac{13}{5} = \dfrac{10}{5} + \dfrac{3}{5} = 2 + \dfrac{3}{5} = 2\dfrac{3}{5}$

1. $\dfrac{13}{5} =$

2. $\dfrac{12}{5} =$

3. $\dfrac{14}{11} =$

4. $\dfrac{26}{7} =$

5. $\dfrac{23}{7} =$

6. $\dfrac{28}{5} =$

7. $\dfrac{37}{11} =$

8. $\dfrac{38}{8} =$

9. $\dfrac{16}{7} =$

10. $\dfrac{24}{5} =$

11. $\dfrac{27}{10} =$

12. $\dfrac{22}{12} =$

13. $\dfrac{40}{7} =$

14. $\dfrac{32}{11} =$

15. $\dfrac{25}{7} =$

16. $\dfrac{22}{8} =$

17. $\dfrac{33}{7} =$

18. $\dfrac{28}{8} =$

19. $\dfrac{45}{10} =$

20. $\dfrac{41}{9} =$

21. $\dfrac{33}{5} =$

22. $\dfrac{33}{14} =$

23. $\dfrac{42}{13} =$

24. $\dfrac{47}{11} =$

Name:

Change the following improper fractions to mixed numbers.

Example:　$\dfrac{13}{5} = \dfrac{10}{5} + \dfrac{3}{5} = 2 + \dfrac{3}{5} = 2\,\dfrac{3}{5}$

1. $\dfrac{13}{2} =$

2. $\dfrac{47}{6} =$

3. $\dfrac{7}{3} =$

4. $\dfrac{39}{7} =$

5. $\dfrac{67}{8} =$

6. $\dfrac{24}{9} =$

7. $\dfrac{23}{4} =$

8. $\dfrac{75}{10} =$

9. $\dfrac{60}{8} =$

10. $\dfrac{20}{7} =$

11. $\dfrac{46}{5} =$

12. $\dfrac{42}{5} =$

13. $\dfrac{27}{8} =$

14. $\dfrac{86}{9} =$

15. $\dfrac{33}{7} =$

16. $\dfrac{9}{6} =$

17. $\dfrac{49}{8} =$

18. $\dfrac{38}{9} =$

19. $\dfrac{60}{9} =$

20. $\dfrac{19}{7} =$

21. $\dfrac{38}{5} =$

22. $\dfrac{37}{5} =$

23. $\dfrac{49}{15} =$

24. $\dfrac{65}{8} =$

Change the following improper fractions to mixed numbers.

Example: $\dfrac{13}{5} = \dfrac{10}{5} + \dfrac{3}{5} = 2 + \dfrac{3}{5} = 2\dfrac{3}{5}$

1. $\dfrac{44}{5} =$ 2. $\dfrac{23}{6} =$ 3. $\dfrac{42}{8} =$

4. $\dfrac{18}{8} =$ 5. $\dfrac{24}{5} =$ 6. $\dfrac{94}{10} =$

7. $\dfrac{47}{6} =$ 8. $\dfrac{70}{6} =$ 9. $\dfrac{17}{3} =$

10. $\dfrac{66}{8} =$ 11. $\dfrac{84}{9} =$ 12. $\dfrac{30}{7} =$

13. $\dfrac{36}{5} =$ 14. $\dfrac{76}{9} =$ 15. $\dfrac{59}{4} =$

16. $\dfrac{22}{8} =$ 17. $\dfrac{54}{7} =$ 18. $\dfrac{28}{9} =$

19. $\dfrac{40}{9} =$ 20. $\dfrac{32}{11} =$ 21. $\dfrac{53}{6} =$

22. $\dfrac{47}{7} =$ 23. $\dfrac{89}{12} =$ 24. $\dfrac{64}{12} =$

Change the following improper fractions to mixed numbers.

Example: $\dfrac{13}{5} = \dfrac{10}{5} + \dfrac{3}{5} = 2 + \dfrac{3}{5} = 2\dfrac{3}{5}$

1. $\dfrac{27}{4} =$

2. $\dfrac{59}{9} =$

3. $\dfrac{73}{8} =$

4. $\dfrac{89}{8} =$

5. $\dfrac{27}{4} =$

6. $\dfrac{61}{5} =$

7. $\dfrac{54}{5} =$

8. $\dfrac{40}{7} =$

9. $\dfrac{110}{12} =$

10. $\dfrac{75}{8} =$

11. $\dfrac{22}{8} =$

12. $\dfrac{69}{9} =$

13. $\dfrac{26}{3} =$

14. $\dfrac{30}{9} =$

15. $\dfrac{47}{8} =$

16. $\dfrac{47}{8} =$

17. $\dfrac{57}{8} =$

18. $\dfrac{102}{11} =$

19. $\dfrac{55}{4} =$

20. $\dfrac{33}{5} =$

21. $\dfrac{54}{7} =$

22. $\dfrac{80}{11} =$

23. $\dfrac{95}{13} =$

24. $\dfrac{123}{15} =$

Change the following improper fractions to mixed numbers.

Example: $\dfrac{13}{5} = \dfrac{10}{5} + \dfrac{3}{5} = 2 + \dfrac{3}{5} = 2\dfrac{3}{5}$

1. $\dfrac{48}{5} =$ 2. $\dfrac{46}{8} =$ 3. $\dfrac{69}{7} =$

4. $\dfrac{23}{7} =$ 5. $\dfrac{33}{6} =$ 6. $\dfrac{42}{5} =$

7. $\dfrac{38}{6} =$ 8. $\dfrac{97}{9} =$ 9. $\dfrac{44}{9} =$

10. $\dfrac{64}{7} =$ 11. $\dfrac{66}{8} =$ 12. $\dfrac{73}{6} =$

13. $\dfrac{30}{4} =$ 14. $\dfrac{39}{7} =$ 15. $\dfrac{113}{9} =$

16. $\dfrac{91}{5} =$ 17. $\dfrac{119}{10} =$ 18. $\dfrac{147}{12} =$

19. $\dfrac{123}{11} =$ 20. $\dfrac{97}{13} =$ 21. $\dfrac{139}{15} =$

22. $\dfrac{107}{15} =$ 23. $\dfrac{68}{5} =$ 24. $\dfrac{198}{20} =$

Name:

Lesson 4-8   Changing improper fractions to mixed numbers

Change the following improper fractions to mixed numbers.

Example: $\dfrac{13}{5} = \dfrac{10}{5} + \dfrac{3}{5} = 2 + \dfrac{3}{5} = 2\,\dfrac{3}{5}$

1. $\dfrac{43}{4} =$ 2. $\dfrac{54}{8} =$ 3. $\dfrac{26}{4} =$

4. $\dfrac{54}{7} =$ 5. $\dfrac{75}{6} =$ 6. $\dfrac{39}{11} =$

7. $\dfrac{37}{3} =$ 8. $\dfrac{80}{9} =$ 9. $\dfrac{64}{9} =$

10. $\dfrac{87}{9} =$ 11. $\dfrac{77}{8} =$ 12. $\dfrac{27}{2} =$

13. $\dfrac{39}{2} =$ 14. $\dfrac{87}{6} =$ 15. $\dfrac{93}{8} =$

16. $\dfrac{109}{7} =$ 17. $\dfrac{97}{10} =$ 18. $\dfrac{72}{14} =$

19. $\dfrac{107}{13} =$ 20. $\dfrac{100}{16} =$ 21. $\dfrac{93}{15} =$

22. $\dfrac{79}{17} =$ 23. $\dfrac{79}{18} =$ 24. $\dfrac{100}{19} =$

Change the following improper fractions to mixed numbers.

Example: $\dfrac{13}{5} = \dfrac{10}{5} + \dfrac{3}{5} = 2 + \dfrac{3}{5} = 2\dfrac{3}{5}$

1. $\dfrac{77}{8} =$

2. $\dfrac{48}{9} =$

3. $\dfrac{60}{7} =$

4. $\dfrac{38}{3} =$

5. $\dfrac{41}{3} =$

6. $\dfrac{50}{3} =$

7. $\dfrac{52}{8} =$

8. $\dfrac{100}{7} =$

9. $\dfrac{67}{9} =$

10. $\dfrac{53}{3} =$

11. $\dfrac{53}{6} =$

12. $\dfrac{51}{4} =$

13. $\dfrac{45}{7} =$

14. $\dfrac{35}{4} =$

15. $\dfrac{90}{8} =$

16. $\dfrac{96}{11} =$

17. $\dfrac{120}{13} =$

18. $\dfrac{115}{14} =$

19. $\dfrac{121}{17} =$

20. $\dfrac{126}{19} =$

21. $\dfrac{127}{16} =$

22. $\dfrac{128}{20} =$

23. $\dfrac{134}{12} =$

24. $\dfrac{155}{18} =$

Change the following improper fractions to mixed numbers.

Example: $\dfrac{13}{5} = \dfrac{10}{5} + \dfrac{3}{5} = 2 + \dfrac{3}{5} = 2\dfrac{3}{5}$

1. $\dfrac{75}{7} =$ 2. $\dfrac{42}{4} =$ 3. $\dfrac{73}{4} =$

4. $\dfrac{30}{4} =$ 5. $\dfrac{55}{8} =$ 6. $\dfrac{44}{3} =$

7. $\dfrac{47}{5} =$ 8. $\dfrac{36}{5} =$ 9. $\dfrac{75}{9} =$

10. $\dfrac{91}{8} =$ 11. $\dfrac{100}{8} =$ 12. $\dfrac{119}{20} =$

13. $\dfrac{89}{7} =$ 14. $\dfrac{69}{7} =$ 15. $\dfrac{41}{7} =$

16. $\dfrac{187}{20} =$ 17. $\dfrac{145}{11} =$ 18. $\dfrac{71}{3} =$

19. $\dfrac{140}{15} =$ 20. $\dfrac{175}{19} =$ 21. $\dfrac{151}{12} =$

22. $\dfrac{155}{17} =$ 23. $\dfrac{110}{17} =$ 24. $\dfrac{130}{16} =$

Add fractions and write the answers in simplest form.

1. $\dfrac{5}{7} + \dfrac{6}{7} =$

2. $\dfrac{4}{9} + \dfrac{10}{9} =$

3. $\dfrac{4}{5} + \dfrac{7}{5} =$

4. $\dfrac{9}{7} + \dfrac{6}{7} =$

5. $\dfrac{6}{10} + \dfrac{7}{10} =$

6. $\dfrac{3}{5} + \dfrac{8}{5} =$

7. $\dfrac{8}{13} + \dfrac{7}{13} =$

8. $\dfrac{12}{10} + \dfrac{13}{10} =$

9. $\dfrac{10}{9} + \dfrac{5}{9} =$

10. $\dfrac{7}{4} + \dfrac{10}{4} =$

11. $\dfrac{7}{6} + \dfrac{8}{6} =$

12. $\dfrac{3}{8} + \dfrac{11}{8} =$

13. $\dfrac{9}{8} + \dfrac{10}{8} =$

14. $\dfrac{15}{13} + \dfrac{12}{13} =$

15. $\dfrac{5}{4} + \dfrac{6}{4} =$

16. $\dfrac{6}{11} + \dfrac{7}{11} =$

17. $\dfrac{13}{11} + \dfrac{12}{11} =$

18. $\dfrac{9}{6} + \dfrac{7}{6} =$

Add fractions and write the answers in simplest form.

1. $\dfrac{4}{5} + \dfrac{7}{5} =$

2. $\dfrac{9}{5} + \dfrac{7}{5} =$

3. $\dfrac{7}{8} + \dfrac{6}{8} =$

4. $\dfrac{6}{7} + \dfrac{9}{7} =$

5. $\dfrac{5}{4} + \dfrac{6}{4} =$

6. $\dfrac{5}{4} + \dfrac{9}{4} =$

7. $\dfrac{7}{3} + \dfrac{4}{3} =$

8. $\dfrac{9}{10} + \dfrac{7}{10} =$

9. $\dfrac{5}{6} + \dfrac{9}{6} =$

10. $\dfrac{11}{9} + \dfrac{15}{9} =$

11. $\dfrac{13}{10} + \dfrac{23}{10} =$

12. $\dfrac{13}{8} + \dfrac{5}{8} =$

13. $\dfrac{9}{20} + \dfrac{13}{20} =$

14. $\dfrac{5}{2} + \dfrac{8}{2} =$

15. $\dfrac{11}{13} + \dfrac{14}{13} =$

16. $\dfrac{13}{3} + \dfrac{12}{3} =$

17. $\dfrac{17}{15} + \dfrac{12}{15} =$

18. $\dfrac{14}{17} + \dfrac{15}{17} =$

Add fractions and write the answers in simplest form.

1. $1\dfrac{2}{5}+\dfrac{6}{5}=$

2. $2\dfrac{2}{3}+\dfrac{5}{3}=$

3. $2\dfrac{5}{7}+1\dfrac{8}{7}=$

4. $1\dfrac{4}{5}+1\dfrac{7}{5}=$

5. $\dfrac{8}{3}+1\dfrac{2}{3}=$

6. $1\dfrac{5}{9}+1\dfrac{10}{9}=$

7. $1\dfrac{7}{6}+1\dfrac{8}{6}=$

8. $2\dfrac{7}{6}+\dfrac{3}{6}=$

9. $2\dfrac{3}{8}+2\dfrac{9}{8}=$

10. $2\dfrac{3}{10}+1\dfrac{13}{10}=$

11. $1\dfrac{5}{9}+1\dfrac{6}{9}=$

12. $\dfrac{9}{8}+2\dfrac{5}{8}=$

13. $\dfrac{12}{10}+2\dfrac{13}{10}=$

14. $1\dfrac{8}{7}+2\dfrac{9}{7}=$

15. $\dfrac{15}{11}+1\dfrac{13}{11}=$

16. $\dfrac{5}{12}+1\dfrac{11}{12}=$

17. $2\dfrac{5}{12}+1\dfrac{13}{12}=$

18. $2\dfrac{3}{15}+2\dfrac{4}{15}=$

Name: _____

Lesson   5-4   Adding   fractions
with the same denominators

Add fractions and write the answers in simplest form.

1. $2\frac{5}{3} + 1\frac{5}{3} =$

2. $2\frac{3}{4} + 1\frac{7}{4} =$

3. $1\frac{3}{7} + 2\frac{8}{7} =$

4. $3\frac{4}{10} + 3\frac{7}{10} =$

5. $1\frac{12}{11} + \frac{14}{11} =$

6. $2\frac{5}{9} + 1\frac{10}{9} =$

7. $3\frac{3}{5} + 2\frac{9}{5} =$

8. $\frac{4}{7} + 3\frac{6}{7} =$

9. $\frac{15}{12} + 3\frac{7}{12} =$

10. $3\frac{3}{8} + \frac{7}{8} =$

11. $1\frac{21}{20} + 3\frac{22}{20} =$

12. $1\frac{9}{20} + 2\frac{14}{20} =$

13. $2\frac{11}{18} + 3\frac{10}{18} =$

14. $1\frac{13}{15} + 1\frac{18}{15} =$

15. $\frac{8}{15} + 3\frac{11}{15} =$

16. $3\frac{12}{13} + 1\frac{12}{13} =$

17. $3\frac{4}{9} + 3\frac{5}{9} =$

18. $2\frac{8}{19} + 3\frac{7}{19} =$

Add fractions and write the answers in simplest form.

1. $3\dfrac{3}{5} + 2\dfrac{4}{5} =$

2. $4\dfrac{7}{6} + 1\dfrac{4}{6} =$

3. $1\dfrac{7}{10} + 3\dfrac{8}{10} =$

4. $3\dfrac{11}{15} + 2\dfrac{6}{15} =$

5. $4\dfrac{10}{9} + 2\dfrac{9}{9} =$

6. $4\dfrac{10}{9} + 3\dfrac{11}{9} =$

7. $3\dfrac{5}{7} + 3\dfrac{6}{7} =$

8. $2\dfrac{7}{17} + 3\dfrac{18}{17} =$

9. $2\dfrac{5}{12} + 3\dfrac{10}{12} =$

10. $4\dfrac{5}{8} + 2\dfrac{9}{8} =$

11. $1\dfrac{11}{25} + 4\dfrac{12}{25} =$

12. $1\dfrac{9}{10} + 4\dfrac{9}{10} =$

13. $3\dfrac{9}{8} + 4\dfrac{11}{8} =$

14. $\dfrac{7}{11} + \dfrac{9}{11} =$

15. $4\dfrac{13}{16} + 2\dfrac{15}{16} =$

16. $4\dfrac{9}{16} + 4\dfrac{10}{16} =$

17. $4\dfrac{17}{21} + 4\dfrac{18}{21} =$

18. $3\dfrac{21}{23} + 4\dfrac{22}{23} =$

53

Add fractions and write the answers in simplest form.

1. $3\dfrac{5}{7} + 1\dfrac{4}{7} =$

2. $2\dfrac{7}{9} + 2\dfrac{10}{9} =$

3. $4\dfrac{4}{9} + 2\dfrac{10}{9} =$

4. $4\dfrac{19}{20} + 1\dfrac{11}{20} =$

5. $3\dfrac{11}{14} + 4\dfrac{13}{14} =$

6. $2\dfrac{5}{7} + 3\dfrac{8}{7} =$

7. $3\dfrac{3}{4} + 2\dfrac{6}{4} =$

8. $3\dfrac{13}{10} + \dfrac{11}{10} =$

9. $2\dfrac{7}{11} + 3\dfrac{9}{11} =$

10. $1\dfrac{13}{16} + 4\dfrac{9}{16} =$

11. $4\dfrac{18}{17} + 3\dfrac{11}{17} =$

12. $3\dfrac{7}{11} + 2\dfrac{9}{11} =$

13. $1\dfrac{21}{25} + 3\dfrac{17}{25} =$

14. $2\dfrac{10}{12} + 4\dfrac{13}{12} =$

15. $\dfrac{10}{21} + 2\dfrac{11}{21} =$

16. $4\dfrac{19}{24} + 3\dfrac{15}{24} =$

17. $2\dfrac{4}{15} + 4\dfrac{13}{15} =$

18. $4\dfrac{12}{18} + 2\dfrac{20}{18} =$

Add fractions and write the answers in simplest form.

1. $3\dfrac{6}{5} + 3\dfrac{3}{5} =$

2. $1\dfrac{6}{7} + 4\dfrac{9}{7} =$

3. $4\dfrac{12}{16} + 2\dfrac{11}{16} =$

4. $4\dfrac{5}{8} + 2\dfrac{7}{8} =$

5. $4\dfrac{7}{15} + 3\dfrac{8}{15} =$

6. $5\dfrac{17}{20} + 3\dfrac{7}{20} =$

7. $5\dfrac{11}{8} + 2\dfrac{10}{8} =$

8. $4\dfrac{11}{14} + 1\dfrac{13}{14} =$

9. $2\dfrac{8}{9} + 4\dfrac{8}{9} =$

10. $2\dfrac{21}{23} + 4\dfrac{12}{23} =$

11. $3\dfrac{10}{24} + \dfrac{17}{24} =$

12. $4\dfrac{9}{15} + 4\dfrac{10}{15} =$

13. $3\dfrac{8}{11} + 4\dfrac{10}{11} =$

14. $3\dfrac{10}{12} + 5\dfrac{6}{12} =$

15. $2\dfrac{21}{25} + 3\dfrac{22}{25} =$

16. $5\dfrac{19}{17} + \dfrac{18}{17} =$

17. $2\dfrac{12}{17} + 5\dfrac{13}{17} =$

18. $5\dfrac{15}{19} + 2\dfrac{14}{19} =$

Name: _____

Add fractions and write the answers in simplest form.

1. $3\dfrac{5}{6} + 2\dfrac{3}{6} =$

2. $2\dfrac{5}{8} + 2\dfrac{7}{8} =$

3. $1\dfrac{7}{14} + 4\dfrac{9}{14} =$

4. $3\dfrac{5}{4} + 4\dfrac{2}{4} =$

5. $3\dfrac{11}{8} + 3\dfrac{7}{8} =$

6. $4\dfrac{9}{13} + 3\dfrac{11}{13} =$

7. $2\dfrac{18}{22} + 3\dfrac{11}{22} =$

8. $3\dfrac{8}{19} + 2\dfrac{15}{19} =$

9. $5\dfrac{4}{9} + \dfrac{10}{9} =$

10. $4\dfrac{15}{14} + 3\dfrac{11}{14} =$

11. $4\dfrac{13}{16} + 2\dfrac{7}{16} =$

12. $1\dfrac{9}{12} + 5\dfrac{9}{12} =$

13. $3\dfrac{9}{20} + 4\dfrac{12}{20} =$

14. $\dfrac{12}{15} + \dfrac{6}{15} =$

15. $2\dfrac{14}{11} + 3\dfrac{13}{11} =$

16. $3\dfrac{8}{13} + 4\dfrac{7}{13} =$

17. $3\dfrac{15}{16} + 5\dfrac{11}{16} =$

18. $5\dfrac{17}{21} + 2\dfrac{19}{21} =$

Add fractions and write the answers in simplest form.

1. $5\dfrac{6}{5} + 1\dfrac{7}{5} =$

2. $2\dfrac{9}{5} + 1\dfrac{13}{5} =$

3. $3\dfrac{4}{9} + 4\dfrac{11}{9} =$

4. $4\dfrac{15}{17} + 3\dfrac{16}{17} =$

5. $4\dfrac{9}{11} + 4\dfrac{8}{11} =$

6. $2\dfrac{5}{8} + 5\dfrac{9}{8} =$

7. $2\dfrac{9}{25} + 3\dfrac{13}{25} =$

8. $4\dfrac{6}{13} + 4\dfrac{9}{13} =$

9. $1\dfrac{10}{13} + 3\dfrac{11}{13} =$

10. $3\dfrac{20}{22} + 4\dfrac{21}{22} =$

11. $3\dfrac{13}{17} + 2\dfrac{14}{17} =$

12. $5\dfrac{13}{18} + 2\dfrac{15}{18} =$

13. $4\dfrac{20}{12} + 3\dfrac{13}{12} =$

14. $1\dfrac{20}{16} + 3\dfrac{17}{16} =$

15. $4\dfrac{17}{20} + 2\dfrac{18}{20} =$

16. $3\dfrac{7}{12} + 2\dfrac{15}{12} =$

17. $3\dfrac{19}{23} + 3\dfrac{20}{23} =$

18. $3\dfrac{7}{24} + 5\dfrac{17}{24} =$

Add fractions and write the answers in simplest form.

1. $2\dfrac{11}{5} + 5\dfrac{12}{5} =$

2. $4\dfrac{7}{8} + 4\dfrac{7}{8} =$

3. $4\dfrac{6}{13} + 2\dfrac{8}{13} =$

4. $1\dfrac{6}{10} + 5\dfrac{9}{10} =$

5. $3\dfrac{8}{7} + 5\dfrac{9}{7} =$

6. $4\dfrac{10}{9} + 3\dfrac{12}{9} =$

7. $5\dfrac{13}{15} + 3\dfrac{14}{15} =$

8. $2\dfrac{7}{16} + 2\dfrac{18}{16} =$

9. $5\dfrac{4}{25} + 2\dfrac{23}{25} =$

10. $4\dfrac{19}{24} + 5\dfrac{17}{24} =$

11. $2\dfrac{12}{18} + 4\dfrac{15}{18} =$

12. $4\dfrac{21}{14} + 3\dfrac{13}{14} =$

13. $5\dfrac{11}{17} + \dfrac{13}{17} =$

14. $3\dfrac{5}{11} + 4\dfrac{9}{11} =$

15. $2\dfrac{20}{23} + 3\dfrac{22}{23} =$

16. $3\dfrac{19}{23} + 3\dfrac{21}{23} =$

17. $4\dfrac{9}{12} + 3\dfrac{12}{12} =$

18. $5\dfrac{24}{17} + 1\dfrac{11}{17} =$

Name:

Subtract fractions and write the answers in simplest form.

1. $1\dfrac{2}{3} - 1\dfrac{1}{3} =$

2. $1\dfrac{2}{9} - \dfrac{7}{9} =$

3. $2\dfrac{3}{7} - \dfrac{5}{7} =$

4. $1\dfrac{3}{10} - \dfrac{9}{10} =$

5. $1\dfrac{1}{5} - \dfrac{4}{5} =$

6. $1\dfrac{3}{5} - \dfrac{2}{5} =$

7. $2\dfrac{2}{11} - 1\dfrac{5}{11} =$

8. $1\dfrac{7}{8} - \dfrac{5}{8} =$

9. $1\dfrac{2}{9} - \dfrac{5}{9} =$

10. $2\dfrac{3}{11} - 1\dfrac{10}{11} =$

11. $2\dfrac{3}{10} - 1\dfrac{1}{10} =$

12. $2\dfrac{4}{7} - 1\dfrac{8}{7} =$

13. $1\dfrac{7}{13} - \dfrac{10}{13} =$

14. $2\dfrac{1}{14} - 1\dfrac{9}{14} =$

15. $2\dfrac{1}{20} - \dfrac{11}{20} =$

16. $2\dfrac{5}{6} - 1\dfrac{7}{6} =$

17. $1\dfrac{12}{15} - \dfrac{14}{15} =$

18. $2\dfrac{3}{17} - 1\dfrac{13}{17} =$

Subtract fractions and write the answers in simplest form.

1. $2 - \dfrac{5}{6} =$

2. $3 - \dfrac{9}{7} =$

3. $\dfrac{21}{8} - \dfrac{14}{8} =$

4. $2\dfrac{3}{5} - \dfrac{7}{5} =$

5. $2\dfrac{3}{10} - 1\dfrac{9}{10} =$

6. $2\dfrac{2}{21} - 1\dfrac{19}{21} =$

7. $3\dfrac{1}{13} - 2\dfrac{9}{13} =$

8. $\dfrac{24}{9} - 1\dfrac{2}{9} =$

9. $1\dfrac{2}{7} - \dfrac{8}{7} =$

10. $3\dfrac{1}{12} - 2\dfrac{5}{12} =$

11. $2\dfrac{11}{17} - \dfrac{9}{17} =$

12. $\dfrac{13}{4} - 2\dfrac{1}{4} =$

13. $1\dfrac{5}{11} - \dfrac{10}{11} =$

14. $2\dfrac{5}{18} - \dfrac{17}{18} =$

15. $3\dfrac{4}{25} - 1\dfrac{23}{25} =$

16. $1\dfrac{7}{23} - \dfrac{2}{23} =$

17. $2\dfrac{7}{19} - \dfrac{20}{19} =$

18. $\dfrac{33}{16} - 1\dfrac{1}{16} =$

Subtract fractions and write the answers in simplest form.

1. $3\dfrac{1}{6} - 1\dfrac{2}{6} =$

2. $3\dfrac{1}{8} - 1\dfrac{3}{8} =$

3. $2\dfrac{1}{7} - \dfrac{5}{7} =$

4. $2\dfrac{2}{9} - \dfrac{11}{9} =$

5. $4 - 2\dfrac{2}{3} =$

6. $2\dfrac{3}{14} - 1\dfrac{11}{14} =$

7. $1\dfrac{3}{14} - \dfrac{11}{14} =$

8. $4\dfrac{1}{13} - 2\dfrac{9}{13} =$

9. $\dfrac{33}{24} - 1\dfrac{1}{24} =$

10. $3 - 1\dfrac{6}{7} =$

11. $3\dfrac{2}{15} - 1\dfrac{7}{15} =$

12. $2\dfrac{3}{11} - \dfrac{13}{11} =$

13. $4\dfrac{3}{23} - 1\dfrac{22}{23} =$

14. $4\dfrac{2}{27} - 3\dfrac{3}{27} =$

15. $\dfrac{40}{18} - \dfrac{16}{18} =$

16. $3\dfrac{4}{17} - 1\dfrac{16}{17} =$

17. $4\dfrac{5}{12} - 2\dfrac{13}{12} =$

18. $4\dfrac{3}{14} - 2\dfrac{15}{14} =$

Subtract fractions and write the answers in simplest form.

1. $4\dfrac{1}{3} - 2\dfrac{2}{3} =$

2. $2\dfrac{3}{5} - 1\dfrac{4}{5} =$

3. $3\dfrac{1}{4} - 1\dfrac{3}{4} =$

4. $4\dfrac{2}{7} - 2\dfrac{6}{7} =$

5. $2\dfrac{5}{13} - \dfrac{9}{13} =$

6. $3\dfrac{1}{8} - 2\dfrac{5}{8} =$

7. $5 - 4\dfrac{5}{16} =$

8. $4\dfrac{5}{14} - 1\dfrac{12}{14} =$

9. $2\dfrac{4}{9} - 1\dfrac{8}{9} =$

10. $3\dfrac{5}{21} - \dfrac{19}{21} =$

11. $5\dfrac{1}{11} - 3\dfrac{2}{11} =$

12. $5\dfrac{7}{20} - 2\dfrac{17}{20} =$

13. $3\dfrac{15}{26} - 1\dfrac{21}{26} =$

14. $5\dfrac{3}{16} - 3\dfrac{15}{16} =$

15. $4\dfrac{7}{15} - 1\dfrac{4}{15} =$

16. $4\dfrac{1}{12} - 2\dfrac{11}{12} =$

17. $3\dfrac{3}{17} - \dfrac{16}{17} =$

18. $5\dfrac{7}{18} - 2\dfrac{13}{18} =$

Lesson 6-5   Subtracting fractions
with the same denominators

Subtract fractions and write the answers in simplest form.

1. $6\frac{1}{4} - 2\frac{3}{4} =$

2. $4\frac{2}{9} - 2\frac{7}{9} =$

3. $4\frac{3}{8} - 3\frac{7}{8} =$

4. $5\frac{1}{5} - 1\frac{4}{5} =$

5. $5\frac{2}{21} - 3\frac{17}{21} =$

6. $3\frac{3}{17} - \frac{15}{17} =$

7. $6\frac{3}{17} - 3\frac{9}{17} =$

8. $5\frac{9}{13} - 2\frac{11}{13} =$

9. $4\frac{11}{25} - 1\frac{23}{25} =$

10. $6\frac{5}{18} - 3\frac{17}{18} =$

11. $3\frac{5}{13} - 2\frac{1}{13} =$

12. $6\frac{3}{16} - 2\frac{13}{16} =$

13. $6\frac{7}{30} - 5\frac{23}{30} =$

14. $5 - 3\frac{15}{19} =$

15. $5\frac{2}{26} - 1\frac{19}{26} =$

16. $4\frac{4}{18} - 1\frac{15}{18} =$

17. $6 - \frac{3}{17} =$

18. $3\frac{2}{23} - \frac{22}{23} =$

Name:

Subtract fractions and write the answers in simplest form.

1. $6 - \dfrac{7}{3} =$

2. $5 - \dfrac{13}{8} =$

3. $4\dfrac{1}{4} - 2\dfrac{2}{4} =$

4. $2\dfrac{2}{7} - \dfrac{5}{7} =$

5. $5\dfrac{3}{9} - 2\dfrac{8}{9} =$

6. $5\dfrac{8}{13} - 1\dfrac{10}{13} =$

7. $3\dfrac{1}{6} - \dfrac{5}{6} =$

8. $4\dfrac{6}{23} - 3\dfrac{19}{23} =$

9. $6\dfrac{7}{13} - 3\dfrac{12}{13} =$

10. $5\dfrac{7}{15} - 2\dfrac{13}{15} =$

11. $4\dfrac{2}{14} - 1\dfrac{5}{14} =$

12. $6\dfrac{1}{20} - 5\dfrac{9}{20} =$

13. $6\dfrac{3}{19} - 4\dfrac{11}{19} =$

14. $4\dfrac{5}{17} - 2\dfrac{15}{17} =$

15. $5\dfrac{6}{17} - 1\dfrac{13}{17} =$

16. $6\dfrac{3}{28} - 1\dfrac{25}{28} =$

17. $6\dfrac{1}{11} - 2\dfrac{8}{11} =$

18. $6\dfrac{5}{18} - 4\dfrac{16}{18} =$

**Lesson 6-7   Subtracting fractions
with the same denominators**

Subtract fractions and write the answers in simplest form.

1. $1\dfrac{1}{6} - \dfrac{5}{6} =$ 2. $3\dfrac{1}{8} - 2\dfrac{7}{8} =$

3. $2\dfrac{3}{7} - 1\dfrac{6}{7} =$ 4. $4\dfrac{2}{9} - \dfrac{8}{9} =$

5. $4\dfrac{2}{15} - 2\dfrac{11}{15} =$ 6. $2\dfrac{3}{13} - \dfrac{12}{13} =$

7. $3\dfrac{5}{14} - \dfrac{11}{14} =$ 8. $3\dfrac{3}{11} - 1\dfrac{9}{11} =$

9. $4\dfrac{5}{24} - 2\dfrac{17}{24} =$ 10. $6\dfrac{5}{14} - 4\dfrac{11}{14} =$

11. $5\dfrac{3}{16} - 3\dfrac{15}{16} =$ 12. $5\dfrac{1}{12} - 1\dfrac{11}{12} =$

13. $6\dfrac{7}{19} - 4\dfrac{18}{19} =$ 14. $4\dfrac{2}{13} - 2\dfrac{12}{13} =$

15. $4\dfrac{3}{23} - \dfrac{11}{23} =$ 16. $4\dfrac{3}{28} - 1\dfrac{19}{28} =$

17. $3\dfrac{9}{25} - 1\dfrac{3}{25} =$ 18. $5\dfrac{4}{19} - 1\dfrac{16}{19} =$

Subtract fractions and write the answers in simplest form.

1. $2\dfrac{5}{8} - \dfrac{7}{8} =$

2. $2\dfrac{1}{7} - 1\dfrac{6}{7} =$

3. $3\dfrac{1}{5} - 2\dfrac{4}{5} =$

4. $4\dfrac{2}{9} - \dfrac{7}{9} =$

5. $4\dfrac{3}{13} - 1\dfrac{9}{13} =$

6. $3\dfrac{5}{14} - 1\dfrac{11}{14} =$

7. $3\dfrac{4}{25} - 1\dfrac{17}{25} =$

8. $6\dfrac{1}{11} - 4\dfrac{9}{11} =$

9. $6\dfrac{3}{17} - 3\dfrac{9}{17} =$

10. $5\dfrac{2}{12} - 4\dfrac{10}{12} =$

11. $5\dfrac{3}{20} - 1\dfrac{14}{20} =$

12. $6\dfrac{3}{19} - 2\dfrac{15}{19} =$

13. $6\dfrac{2}{17} - 5\dfrac{5}{17} =$

14. $5\dfrac{3}{16} - 1\dfrac{7}{16} =$

15. $4\dfrac{6}{11} - 2\dfrac{10}{11} =$

16. $4\dfrac{5}{23} - 2\dfrac{19}{23} =$

17. $5\dfrac{2}{12} - 2\dfrac{10}{12} =$

18. $6\dfrac{2}{15} - 3\dfrac{13}{15} =$

Subtract fractions and write the answers in simplest form.

1. $3\dfrac{1}{4} - \dfrac{5}{4} =$

2. $4\dfrac{1}{9} - 3\dfrac{7}{9} =$

3. $4\dfrac{1}{8} - 2\dfrac{9}{8} =$

4. $7\dfrac{3}{5} - 4\dfrac{4}{5} =$

5. $2\dfrac{3}{17} - 1\dfrac{19}{17} =$

6. $5\dfrac{3}{20} - 1\dfrac{13}{20} =$

7. $4\dfrac{5}{14} - \dfrac{9}{14} =$

8. $4\dfrac{5}{17} - 2\dfrac{14}{17} =$

9. $5\dfrac{9}{25} - 1\dfrac{17}{25} =$

10. $6\dfrac{2}{19} - 3\dfrac{20}{19} =$

11. $6\dfrac{7}{21} - \dfrac{17}{21} =$

12. $5\dfrac{6}{13} - 2\dfrac{8}{13} =$

13. $7\dfrac{5}{14} - 3\dfrac{13}{14} =$

14. $6\dfrac{4}{25} - 4\dfrac{21}{25} =$

15. $6\dfrac{4}{30} - 4\dfrac{21}{30} =$

16. $5\dfrac{6}{15} - 2\dfrac{13}{15} =$

17. $5\dfrac{3}{16} - 1\dfrac{19}{16} =$

18. $7\dfrac{5}{12} - 5\dfrac{11}{12} =$

Subtract fractions and write the answers in simplest form.

1. $5\dfrac{3}{5} - 1\dfrac{8}{5} =$

2. $7\dfrac{1}{9} - 3\dfrac{8}{9} =$

3. $6\dfrac{5}{6} - \dfrac{9}{6} =$

4. $6\dfrac{3}{7} - 3\dfrac{10}{7} =$

5. $3\dfrac{3}{26} - 1\dfrac{15}{26} =$

6. $4\dfrac{2}{11} - 1\dfrac{8}{11} =$

7. $4\dfrac{2}{17} - 2\dfrac{14}{17} =$

8. $5\dfrac{4}{15} - 2\dfrac{11}{15} =$

9. $6\dfrac{4}{21} - 2\dfrac{22}{21} =$

10. $5\dfrac{3}{16} - 4\dfrac{13}{16} =$

11. $5\dfrac{5}{12} - 1\dfrac{17}{12} =$

12. $4\dfrac{5}{17} - 1\dfrac{14}{17} =$

13. $4\dfrac{2}{19} - 3\dfrac{9}{19} =$

14. $6\dfrac{6}{25} - 2\dfrac{16}{25} =$

15. $5\dfrac{3}{16} - 2\dfrac{18}{16} =$

16. $5\dfrac{2}{13} - 4\dfrac{9}{13} =$

17. $4\dfrac{4}{30} - 1\dfrac{21}{30} =$

18. $6\dfrac{5}{12} - 2\dfrac{7}{12} =$

Add fractions and write the answers in simplest form.

1. $\dfrac{1}{2} + \dfrac{1}{4} =$

2. $\dfrac{2}{5} + \dfrac{3}{10} =$

3. $\dfrac{1}{2} + \dfrac{1}{6} =$

4. $\dfrac{1}{6} + \dfrac{5}{12} =$

5. $\dfrac{2}{3} + \dfrac{1}{6} =$

6. $\dfrac{3}{4} + \dfrac{7}{12} =$

7. $\dfrac{3}{5} + \dfrac{1}{10} =$

8. $\dfrac{2}{7} + \dfrac{5}{14} =$

9. $\dfrac{2}{7} + \dfrac{3}{14} =$

10. $\dfrac{2}{3} + \dfrac{7}{9} =$

11. $\dfrac{1}{2} + \dfrac{3}{8} =$

12. $\dfrac{1}{4} + \dfrac{5}{8} =$

13. $\dfrac{2}{3} + \dfrac{4}{9} =$

14. $\dfrac{1}{2} + \dfrac{7}{8} =$

15. $\dfrac{1}{2} + \dfrac{7}{10} =$

16. $\dfrac{2}{5} + \dfrac{11}{15} =$

17. $\dfrac{3}{4} + \dfrac{7}{8} =$

18. $\dfrac{3}{8} + \dfrac{15}{16} =$

Name: _____

Add fractions and write the answers in simplest form.

1. $\dfrac{3}{4} + \dfrac{5}{8} =$

2. $\dfrac{2}{3} + \dfrac{5}{9} =$

3. $\dfrac{2}{5} + \dfrac{3}{10} =$

4. $\dfrac{1}{2} + \dfrac{7}{12} =$

5. $\dfrac{1}{4} + \dfrac{7}{12} =$

6. $\dfrac{1}{5} + \dfrac{6}{15} =$

7. $\dfrac{5}{6} + \dfrac{7}{18} =$

8. $\dfrac{5}{12} + \dfrac{7}{24} =$

9. $\dfrac{2}{7} + \dfrac{3}{14} =$

10. $\dfrac{1}{7} + \dfrac{11}{21} =$

11. $\dfrac{4}{5} + \dfrac{11}{15} =$

12. $\dfrac{2}{5} + \dfrac{9}{10} =$

13. $\dfrac{5}{9} + \dfrac{13}{18} =$

14. $\dfrac{1}{6} + \dfrac{13}{18} =$

15. $\dfrac{3}{10} + \dfrac{9}{20} =$

16. $\dfrac{4}{5} + \dfrac{3}{20} =$

17. $\dfrac{2}{5} + \dfrac{7}{20} =$

18. $\dfrac{1}{4} + \dfrac{15}{16} =$

Name: _____

Add fractions and write the answers in simplest form.

1. $\dfrac{1}{2} + \dfrac{1}{3} =$

2. $\dfrac{1}{3} + \dfrac{2}{5} =$

3. $\dfrac{2}{3} + \dfrac{1}{4} =$

4. $\dfrac{2}{3} + \dfrac{3}{4} =$

5. $\dfrac{1}{3} + \dfrac{1}{4} =$

6. $\dfrac{1}{2} + \dfrac{4}{5} =$

7. $\dfrac{2}{3} + \dfrac{1}{6} =$

8. $\dfrac{1}{3} + \dfrac{7}{9} =$

9. $\dfrac{2}{3} + \dfrac{2}{5} =$

10. $\dfrac{1}{4} + \dfrac{2}{5} =$

11. $\dfrac{3}{10} + \dfrac{4}{20} =$

12. $\dfrac{3}{4} + \dfrac{7}{16} =$

13. $\dfrac{3}{4} + \dfrac{2}{5} =$

14. $\dfrac{2}{7} + \dfrac{5}{14} =$

15. $\dfrac{1}{3} + \dfrac{7}{15} =$

16. $\dfrac{1}{4} + \dfrac{5}{6} =$

17. $\dfrac{2}{5} + \dfrac{7}{15} =$

18. $\dfrac{2}{3} + \dfrac{11}{12} =$

Add fractions and write the answers in simplest form.

1. $\dfrac{1}{2} + \dfrac{1}{5} =$

2. $\dfrac{1}{2} + \dfrac{3}{7} =$

3. $\dfrac{2}{3} + \dfrac{5}{9} =$

4. $\dfrac{2}{3} + \dfrac{2}{7} =$

5. $\dfrac{1}{2} + \dfrac{2}{7} =$

6. $\dfrac{1}{5} + \dfrac{11}{20} =$

7. $\dfrac{2}{3} + \dfrac{3}{4} =$

8. $\dfrac{5}{9} + \dfrac{7}{18} =$

9. $\dfrac{1}{2} + \dfrac{9}{14} =$

10. $\dfrac{5}{6} + \dfrac{2}{5} =$

11. $\dfrac{1}{3} + \dfrac{2}{5} =$

12. $\dfrac{2}{3} + \dfrac{3}{4} =$

13. $\dfrac{3}{4} + \dfrac{1}{6} =$

14. $\dfrac{3}{11} + \dfrac{5}{22} =$

15. $\dfrac{1}{3} + \dfrac{5}{6} =$

16. $\dfrac{1}{4} + \dfrac{6}{7} =$

17. $\dfrac{1}{4} + \dfrac{2}{5} =$

18. $\dfrac{2}{3} + \dfrac{7}{8} =$

Add fractions and write the answers in simplest form.

1. $\dfrac{1}{2} + \dfrac{1}{3} =$

2. $\dfrac{2}{7} + \dfrac{5}{21} =$

3. $\dfrac{2}{3} + \dfrac{1}{4} =$

4. $\dfrac{1}{8} + \dfrac{2}{5} =$

5. $\dfrac{1}{2} + \dfrac{4}{9} =$

6. $\dfrac{7}{12} + \dfrac{7}{36} =$

7. $\dfrac{1}{3} + \dfrac{5}{7} =$

8. $\dfrac{3}{8} + \dfrac{5}{6} =$

9. $\dfrac{2}{3} + \dfrac{1}{15} =$

10. $\dfrac{2}{9} + \dfrac{2}{5} =$

11. $\dfrac{3}{11} + \dfrac{7}{22} =$

12. $\dfrac{1}{6} + \dfrac{13}{24} =$

13. $\dfrac{2}{7} + \dfrac{1}{6} =$

14. $\dfrac{5}{7} + \dfrac{3}{5} =$

15. $\dfrac{3}{4} + \dfrac{2}{5} =$

16. $\dfrac{5}{12} + \dfrac{5}{18} =$

17. $\dfrac{4}{15} + \dfrac{11}{30} =$

18. $\dfrac{2}{13} + \dfrac{9}{39} =$

Add fractions and write the answers in simplest form.

1. $\dfrac{1}{3} + \dfrac{3}{10} =$

2. $\dfrac{6}{13} + \dfrac{7}{26} =$

3. $\dfrac{1}{2} + \dfrac{4}{9} =$

4. $\dfrac{4}{5} + \dfrac{3}{7} =$

5. $\dfrac{3}{4} + \dfrac{3}{5} =$

6. $\dfrac{3}{10} + \dfrac{17}{40} =$

7. $\dfrac{2}{5} + \dfrac{9}{25} =$

8. $\dfrac{5}{6} + \dfrac{3}{7} =$

9. $\dfrac{7}{11} + \dfrac{7}{33} =$

10. $\dfrac{2}{5} + \dfrac{4}{9} =$

11. $\dfrac{2}{7} + \dfrac{11}{28} =$

12. $\dfrac{5}{11} + \dfrac{3}{44} =$

13. $\dfrac{5}{9} + \dfrac{14}{27} =$

14. $\dfrac{7}{12} + \dfrac{4}{15} =$

15. $\dfrac{5}{12} + \dfrac{13}{18} =$

16. $\dfrac{3}{10} + \dfrac{5}{12} =$

17. $\dfrac{3}{5} + \dfrac{4}{9} =$

18. $\dfrac{1}{6} + \dfrac{23}{36} =$

Add fractions and write the answers in simplest form.

1. $\dfrac{1}{3} + \dfrac{3}{5} =$

2. $\dfrac{7}{12} + \dfrac{13}{48} =$

3. $\dfrac{3}{11} + \dfrac{10}{33} =$

4. $\dfrac{2}{9} + \dfrac{23}{45} =$

5. $\dfrac{3}{4} + \dfrac{2}{7} =$

6. $\dfrac{7}{16} + \dfrac{13}{24} =$

7. $\dfrac{5}{6} + \dfrac{4}{9} =$

8. $\dfrac{7}{15} + \dfrac{26}{45} =$

9. $\dfrac{3}{5} + \dfrac{3}{4} =$

10. $\dfrac{4}{9} + \dfrac{7}{8} =$

11. $\dfrac{3}{14} + \dfrac{9}{42} =$

12. $\dfrac{3}{5} + \dfrac{5}{6} =$

13. $\dfrac{7}{12} + \dfrac{23}{60} =$

14. $\dfrac{7}{12} + \dfrac{5}{18} =$

15. $\dfrac{3}{4} + \dfrac{8}{9} =$

16. $\dfrac{9}{14} + \dfrac{2}{35} =$

17. $\dfrac{6}{7} + \dfrac{7}{8} =$

18. $\dfrac{5}{18} + \dfrac{22}{45} =$

Add fractions and write the answers in simplest form.

1. $\dfrac{1}{2} + \dfrac{3}{7} =$

2. $\dfrac{4}{5} + \dfrac{3}{4} =$

3. $\dfrac{2}{3} + \dfrac{3}{8} =$

4. $\dfrac{2}{7} + \dfrac{5}{8} =$

5. $\dfrac{1}{2} + \dfrac{7}{11} =$

6. $\dfrac{3}{5} + \dfrac{5}{7} =$

7. $\dfrac{1}{3} + \dfrac{10}{21} =$

8. $\dfrac{1}{4} + \dfrac{2}{3} =$

9. $\dfrac{7}{12} + \dfrac{11}{48} =$

10. $\dfrac{9}{14} + \dfrac{11}{21} =$

11. $\dfrac{5}{12} + \dfrac{7}{9} =$

12. $\dfrac{7}{15} + \dfrac{22}{45} =$

13. $\dfrac{3}{8} + \dfrac{11}{12} =$

14. $\dfrac{5}{16} + \dfrac{5}{24} =$

15. $\dfrac{4}{5} + \dfrac{6}{7} =$

16. $\dfrac{7}{12} + \dfrac{13}{18} =$

17. $\dfrac{5}{6} + \dfrac{8}{9} =$

18. $\dfrac{13}{18} + \dfrac{4}{45} =$

Add fractions and write the answers in simplest form.

1. $1\dfrac{1}{2} + \dfrac{3}{7} =$

2. $\dfrac{7}{12} + 1\dfrac{5}{9} =$

3. $1\dfrac{1}{3} + \dfrac{4}{5} =$

4. $1\dfrac{1}{18} + \dfrac{7}{12} =$

5. $\dfrac{2}{3} + 1\dfrac{3}{14} =$

6. $1\dfrac{3}{14} + \dfrac{22}{35} =$

7. $\dfrac{3}{4} + 1\dfrac{2}{9} =$

8. $\dfrac{5}{16} + 1\dfrac{13}{40} =$

9. $\dfrac{4}{5} + \dfrac{5}{9} =$

10. $1\dfrac{5}{6} + 1\dfrac{3}{7} =$

11. $1\dfrac{1}{12} + 1\dfrac{1}{36} =$

12. $1\dfrac{6}{15} + 1\dfrac{23}{60} =$

13. $1\dfrac{3}{13} + 1\dfrac{7}{52} =$

14. $\dfrac{3}{10} + 1\dfrac{9}{14} =$

15. $\dfrac{7}{8} + 1\dfrac{11}{12} =$

16. $1\dfrac{3}{18} + \dfrac{17}{24} =$

17. $1\dfrac{5}{12} + \dfrac{9}{32} =$

18. $1\dfrac{11}{18} + 1\dfrac{5}{27} =$

Name:

Add fractions and write the answers in simplest form.

1. $1\dfrac{1}{2} + 1\dfrac{3}{11} =$

2. $\dfrac{4}{5} + 1\dfrac{2}{9} =$

3. $\dfrac{3}{4} + 1\dfrac{4}{9} =$

4. $1\dfrac{6}{7} + \dfrac{4}{9} =$

5. $1\dfrac{1}{3} + \dfrac{5}{7} =$

6. $1\dfrac{2}{3} + 1\dfrac{15}{16} =$

7. $1\dfrac{3}{5} + 1\dfrac{3}{8} =$

8. $1\dfrac{3}{4} + \dfrac{16}{21} =$

9. $1\dfrac{1}{6} + \dfrac{8}{9} =$

10. $\dfrac{9}{14} + 1\dfrac{22}{56} =$

11. $\dfrac{5}{12} + 1\dfrac{23}{60} =$

12. $1\dfrac{13}{26} + 1\dfrac{24}{39} =$

13. $1\dfrac{7}{22} + 1\dfrac{12}{33} =$

14. $1\dfrac{7}{8} + 1\dfrac{11}{12} =$

15. $1\dfrac{7}{13} + 1\dfrac{10}{65} =$

16. $1\dfrac{7}{8} + \dfrac{9}{14} =$

17. $\dfrac{5}{8} + 1\dfrac{9}{14} =$

18. $\dfrac{11}{18} + 1\dfrac{13}{27} =$

Name: _____

Subtract fractions and write the answers in simplest form.

1. $\dfrac{1}{2} - \dfrac{1}{3} =$

2. $\dfrac{2}{3} - \dfrac{1}{4} =$

3. $\dfrac{1}{2} - \dfrac{2}{5} =$

4. $\dfrac{1}{3} - \dfrac{1}{5} =$

5. $\dfrac{1}{2} - \dfrac{3}{10} =$

6. $\dfrac{2}{3} - \dfrac{1}{6} =$

7. $\dfrac{3}{4} - \dfrac{3}{8} =$

8. $\dfrac{2}{3} - \dfrac{1}{7} =$

9. $\dfrac{1}{4} - \dfrac{1}{12} =$

10. $\dfrac{2}{5} - \dfrac{1}{10} =$

11. $\dfrac{5}{6} - \dfrac{7}{12} =$

12. $\dfrac{4}{5} - \dfrac{7}{15} =$

13. $\dfrac{5}{6} - \dfrac{5}{8} =$

14. $\dfrac{3}{5} - \dfrac{1}{6} =$

15. $\dfrac{6}{7} - \dfrac{9}{14} =$

16. $\dfrac{1}{5} - \dfrac{1}{8} =$

17. $\dfrac{3}{7} - \dfrac{8}{21} =$

18. $\dfrac{7}{8} - \dfrac{7}{12} =$

Subtract fractions and write the answers in simplest form.

1. $\dfrac{1}{2} - \dfrac{1}{5} =$

2. $\dfrac{2}{3} - \dfrac{3}{7} =$

3. $\dfrac{2}{5} - \dfrac{1}{6} =$

4. $\dfrac{3}{4} - \dfrac{1}{6} =$

5. $\dfrac{2}{3} - \dfrac{1}{4} =$

6. $\dfrac{7}{9} - \dfrac{5}{18} =$

7. $\dfrac{5}{8} - \dfrac{5}{12} =$

8. $\dfrac{7}{10} - \dfrac{11}{30} =$

9. $\dfrac{3}{4} - \dfrac{3}{7} =$

10. $\dfrac{3}{4} - \dfrac{5}{16} =$

11. $\dfrac{3}{4} - \dfrac{1}{6} =$

12. $\dfrac{4}{5} - \dfrac{5}{7} =$

13. $\dfrac{4}{5} - \dfrac{9}{20} =$

14. $\dfrac{5}{6} - \dfrac{3}{7} =$

15. $\dfrac{5}{6} - \dfrac{11}{18} =$

16. $\dfrac{3}{7} - \dfrac{2}{9} =$

17. $\dfrac{7}{9} - \dfrac{5}{12} =$

18. $\dfrac{7}{8} - \dfrac{2}{9} =$

Subtract fractions and write the answers in simplest form.

1. $\dfrac{1}{2} - \dfrac{2}{7} =$

2. $\dfrac{3}{4} - \dfrac{3}{10} =$

3. $\dfrac{3}{4} - \dfrac{2}{5} =$

4. $\dfrac{4}{5} - \dfrac{2}{9} =$

5. $\dfrac{2}{3} - \dfrac{3}{7} =$

6. $\dfrac{11}{13} - \dfrac{9}{39} =$

7. $\dfrac{5}{6} - \dfrac{2}{9} =$

8. $\dfrac{11}{12} - \dfrac{7}{16} =$

9. $\dfrac{1}{8} - \dfrac{1}{24} =$

10. $\dfrac{9}{14} - \dfrac{8}{35} =$

11. $\dfrac{3}{4} - \dfrac{2}{9} =$

12. $\dfrac{4}{15} - \dfrac{1}{30} =$

13. $\dfrac{3}{11} - \dfrac{1}{33} =$

14. $\dfrac{15}{16} - \dfrac{9}{32} =$

15. $\dfrac{7}{12} - \dfrac{5}{18} =$

16. $\dfrac{9}{14} - \dfrac{8}{21} =$

17. $\dfrac{7}{9} - \dfrac{1}{12} =$

18. $\dfrac{15}{16} - \dfrac{13}{24} =$

Subtract fractions and write the answers in simplest form.

1. $\dfrac{2}{3} - \dfrac{4}{9} =$

2. $\dfrac{5}{6} - \dfrac{7}{24} =$

3. $\dfrac{1}{6} - \dfrac{1}{9} =$

4. $\dfrac{2}{3} - \dfrac{1}{8} =$

5. $\dfrac{7}{8} - \dfrac{13}{32} =$

6. $\dfrac{7}{9} - \dfrac{5}{12} =$

7. $\dfrac{5}{6} - \dfrac{9}{14} =$

8. $\dfrac{9}{10} - \dfrac{11}{25} =$

9. $\dfrac{1}{2} - \dfrac{1}{7} =$

10. $\dfrac{9}{11} - \dfrac{25}{44} =$

11. $\dfrac{11}{12} - \dfrac{9}{16} =$

12. $\dfrac{13}{18} - \dfrac{11}{24} =$

13. $\dfrac{5}{8} - \dfrac{1}{9} =$

14. $\dfrac{15}{16} - \dfrac{21}{40} =$

15. $\dfrac{11}{12} - \dfrac{8}{9} =$

16. $\dfrac{7}{12} - \dfrac{17}{60} =$

17. $\dfrac{13}{18} - \dfrac{13}{27} =$

18. $\dfrac{13}{24} - \dfrac{11}{32} =$

Subtract fractions and write the answers in simplest form.

1. $\dfrac{3}{5} - \dfrac{1}{6} =$

2. $\dfrac{3}{4} - \dfrac{3}{7} =$

3. $\dfrac{1}{2} - \dfrac{5}{11} =$

4. $\dfrac{1}{4} - \dfrac{1}{14} =$

5. $\dfrac{5}{6} - \dfrac{4}{9} =$

6. $\dfrac{5}{6} - \dfrac{11}{18} =$

7. $\dfrac{5}{6} - \dfrac{8}{21} =$

8. $\dfrac{13}{15} - \dfrac{11}{18} =$

9. $\dfrac{3}{4} - \dfrac{11}{16} =$

10. $\dfrac{5}{8} - \dfrac{7}{24} =$

11. $\dfrac{3}{8} - \dfrac{1}{20} =$

12. $\dfrac{7}{8} - \dfrac{15}{28} =$

13. $\dfrac{11}{12} - \dfrac{37}{72} =$

14. $\dfrac{10}{13} - \dfrac{31}{52} =$

15. $\dfrac{10}{13} - \dfrac{19}{78} =$

16. $\dfrac{13}{16} - \dfrac{51}{96} =$

17. $\dfrac{14}{15} - \dfrac{28}{45} =$

18. $\dfrac{21}{26} - \dfrac{19}{39} =$

Subtract fractions and write the answers in simplest form.

1. $\dfrac{2}{3} - \dfrac{11}{21} =$

2. $\dfrac{3}{4} - \dfrac{9}{20} =$

3. $\dfrac{3}{4} - \dfrac{3}{5} =$

4. $\dfrac{2}{3} - \dfrac{5}{8} =$

5. $\dfrac{2}{3} - \dfrac{1}{6} =$

6. $\dfrac{4}{5} - \dfrac{16}{25} =$

7. $\dfrac{11}{12} - \dfrac{13}{36} =$

8. $\dfrac{5}{6} - \dfrac{7}{9} =$

9. $\dfrac{11}{15} - \dfrac{17}{45} =$

10. $\dfrac{9}{11} - \dfrac{31}{44} =$

11. $\dfrac{7}{24} - \dfrac{1}{72} =$

12. $\dfrac{12}{14} - \dfrac{13}{35} =$

13. $\dfrac{5}{7} - \dfrac{4}{9} =$

14. $\dfrac{9}{14} - \dfrac{10}{21} =$

15. $\dfrac{13}{16} - \dfrac{7}{24} =$

16. $\dfrac{11}{15} - \dfrac{31}{75} =$

17. $\dfrac{10}{12} - \dfrac{3}{42} =$

18. $\dfrac{13}{18} - \dfrac{23}{45} =$

Subtract fractions and write the answers in simplest form.

1. $\dfrac{1}{8} - \dfrac{1}{10} =$

2. $\dfrac{1}{4} - \dfrac{3}{18} =$

3. $\dfrac{1}{2} - \dfrac{3}{7} =$

4. $\dfrac{2}{3} - \dfrac{3}{7} =$

5. $\dfrac{5}{6} - \dfrac{5}{27} =$

6. $\dfrac{11}{12} - \dfrac{13}{21} =$

7. $\dfrac{3}{4} - \dfrac{5}{9} =$

8. $\dfrac{11}{12} - \dfrac{13}{72} =$

9. $\dfrac{1}{4} - \dfrac{4}{17} =$

10. $\dfrac{7}{8} - \dfrac{4}{9} =$

11. $\dfrac{7}{8} - \dfrac{13}{36} =$

12. $\dfrac{13}{18} - \dfrac{17}{30} =$

13. $\dfrac{10}{12} - \dfrac{20}{48} =$

14. $\dfrac{17}{26} - \dfrac{14}{39} =$

15. $\dfrac{7}{9} - \dfrac{35}{72} =$

16. $\dfrac{7}{8} - \dfrac{15}{28} =$

17. $\dfrac{9}{11} - \dfrac{27}{55} =$

18. $\dfrac{9}{16} - \dfrac{11}{24} =$

Subtract fractions and write the answers in simplest form.

1. $\dfrac{1}{3} - \dfrac{3}{13} =$

2. $\dfrac{5}{9} - \dfrac{5}{12} =$

3. $\dfrac{1}{2} - \dfrac{4}{15} =$

4. $\dfrac{7}{8} - \dfrac{5}{12} =$

5. $\dfrac{5}{7} - \dfrac{3}{10} =$

6. $\dfrac{2}{3} - \dfrac{1}{8} =$

7. $\dfrac{9}{10} - \dfrac{5}{14} =$

8. $\dfrac{7}{13} - \dfrac{1}{26} =$

9. $\dfrac{3}{4} - \dfrac{5}{12} =$

10. $\dfrac{5}{12} - \dfrac{3}{32} =$

11. $\dfrac{7}{8} - \dfrac{5}{12} =$

12. $\dfrac{7}{9} - \dfrac{19}{27} =$

13. $\dfrac{11}{12} - \dfrac{13}{18} =$

14. $\dfrac{11}{12} - \dfrac{9}{84} =$

15. $\dfrac{13}{14} - \dfrac{27}{42} =$

16. $\dfrac{9}{14} - \dfrac{3}{56} =$

17. $\dfrac{9}{16} - \dfrac{7}{40} =$

18. $\dfrac{13}{18} - \dfrac{11}{27} =$

Subtract fractions and write the answers in simplest form.

1. $1\dfrac{1}{2} - \dfrac{1}{3} =$

2. $1\dfrac{1}{4} - \dfrac{5}{6} =$

3. $1\dfrac{2}{3} - \dfrac{7}{9} =$

4. $1\dfrac{1}{3} - \dfrac{3}{4} =$

5. $1\dfrac{1}{6} - \dfrac{7}{12} =$

6. $1\dfrac{1}{6} - \dfrac{11}{18} =$

7. $1\dfrac{1}{3} - \dfrac{3}{4} =$

8. $1\dfrac{2}{5} - \dfrac{13}{20} =$

9. $1\dfrac{2}{3} - \dfrac{4}{5} =$

10. $1\dfrac{1}{5} - \dfrac{5}{6} =$

11. $1\dfrac{3}{8} - \dfrac{11}{12} =$

12. $1\dfrac{1}{6} - \dfrac{7}{8} =$

13. $1\dfrac{1}{2} - \dfrac{6}{7} =$

14. $1\dfrac{3}{7} - \dfrac{8}{9} =$

15. $1\dfrac{1}{4} - \dfrac{4}{5} =$

16. $1\dfrac{2}{13} - \dfrac{15}{26} =$

17. $1\dfrac{1}{6} - \dfrac{3}{5} =$

18. $1\dfrac{1}{10} - \dfrac{13}{30} =$

Subtract fractions and write the answers in simplest form.

1. $1\dfrac{1}{2} - \dfrac{3}{5} =$

2. $3\dfrac{1}{2} - 1\dfrac{6}{7} =$

3. $2\dfrac{1}{4} - 1\dfrac{5}{6} =$

4. $2\dfrac{1}{12} - \dfrac{8}{15} =$

5. $2\dfrac{2}{5} - \dfrac{6}{7} =$

6. $2\dfrac{1}{6} - 1\dfrac{7}{9} =$

7. $1\dfrac{1}{8} - \dfrac{11}{12} =$

8. $3\dfrac{1}{4} - \dfrac{5}{7} =$

9. $2\dfrac{3}{8} - 1\dfrac{7}{9} =$

10. $3\dfrac{2}{15} - 2\dfrac{22}{45} =$

11. $2\dfrac{1}{14} - \dfrac{15}{28} =$

12. $2\dfrac{2}{13} - 1\dfrac{16}{65} =$

13. $1\dfrac{3}{16} - \dfrac{17}{20} =$

14. $3\dfrac{1}{4} - 2\dfrac{9}{10} =$

15. $2\dfrac{3}{11} - \dfrac{25}{44} =$

16. $1\dfrac{5}{12} - \dfrac{17}{18} =$

17. $1\dfrac{2}{9} - \dfrac{16}{21} =$

18. $2\dfrac{1}{14} - 1\dfrac{22}{35} =$

Multiply fractions and write the answers in simplest form.

1. $1\dfrac{1}{2} \times \dfrac{1}{3} =$

2. $\dfrac{1}{3} \times 1\dfrac{2}{3} =$

3. $1\dfrac{1}{4} \times \dfrac{2}{3} =$

4. $\dfrac{2}{3} \times 1\dfrac{3}{4} =$

5. $1\dfrac{2}{3} \times 1\dfrac{1}{5} =$

6. $1\dfrac{2}{3} \times \dfrac{2}{5} =$

7. $\dfrac{3}{4} \times 1\dfrac{2}{5} =$

8. $1\dfrac{1}{4} \times \dfrac{3}{4} =$

9. $\dfrac{2}{7} \times 1\dfrac{3}{5} =$

10. $1\dfrac{3}{5} \times 1\dfrac{2}{7} =$

11. $1\dfrac{2}{3} \times 1\dfrac{5}{6} =$

12. $1\dfrac{3}{7} \times 1\dfrac{4}{5} =$

13. $1\dfrac{1}{4} \times 1\dfrac{2}{3} =$

14. $2\dfrac{2}{9} \times 2\dfrac{3}{8} =$

15. $2\dfrac{1}{5} \times 1\dfrac{5}{6} =$

16. $1\dfrac{1}{5} \times 2\dfrac{5}{8} =$

17. $1\dfrac{3}{4} \times 2\dfrac{2}{3} =$

18. $2\dfrac{4}{5} \times 1\dfrac{3}{4} =$

Multiply fractions and write the answers in simplest form.

1. $1\dfrac{2}{3} \times 1\dfrac{1}{4} =$

2. $1\dfrac{1}{2} \times \dfrac{5}{7} =$

3. $\dfrac{1}{4} \times 1\dfrac{2}{5} =$

4. $\dfrac{2}{3} \times 1\dfrac{4}{5} =$

5. $1\dfrac{3}{4} \times \dfrac{1}{6} =$

6. $2\dfrac{1}{2} \times 2\dfrac{1}{5} =$

7. $2\dfrac{1}{3} \times 1\dfrac{3}{5} =$

8. $1\dfrac{2}{3} \times 2\dfrac{5}{6} =$

9. $2\dfrac{1}{2} \times \dfrac{3}{7} =$

10. $\dfrac{3}{4} \times 1\dfrac{4}{5} =$

11. $\dfrac{2}{3} \times 1\dfrac{5}{8} =$

12. $2\dfrac{5}{6} \times \dfrac{3}{4} =$

13. $1\dfrac{1}{4} \times \dfrac{4}{5} =$

14. $1\dfrac{2}{3} \times 1\dfrac{5}{7} =$

15. $\dfrac{5}{6} \times 2\dfrac{3}{7} =$

16. $2\dfrac{1}{4} \times 1\dfrac{6}{7} =$

17. $2\dfrac{3}{4} \times 2\dfrac{5}{8} =$

18. $1\dfrac{2}{5} \times 2\dfrac{4}{5} =$

Multiply fractions and write the answers in simplest form.

1. $\dfrac{5}{3} \times 1\dfrac{2}{3} =$ 2. $1\dfrac{1}{4} \times \dfrac{8}{3} =$

3. $\dfrac{7}{2} \times 1\dfrac{3}{4} =$ 4. $\dfrac{10}{3} \times 1\dfrac{3}{5} =$

5. $1\dfrac{2}{3} \times \dfrac{5}{4} =$ 6. $1\dfrac{2}{3} \times 1\dfrac{7}{8} =$

7. $1\dfrac{1}{3} \times \dfrac{3}{7} =$ 8. $2\dfrac{5}{4} \times 2\dfrac{2}{5} =$

9. $2\dfrac{2}{5} \times 1\dfrac{7}{6} =$ 10. $1\dfrac{3}{5} \times 2\dfrac{6}{7} =$

11. $1\dfrac{3}{4} \times 2\dfrac{5}{8} =$ 12. $2\dfrac{9}{8} \times 1\dfrac{2}{9} =$

13. $2\dfrac{2}{3} \times 2\dfrac{4}{9} =$ 14. $1\dfrac{3}{5} \times 1\dfrac{2}{9} =$

15. $2\dfrac{5}{4} \times \dfrac{6}{7} =$ 16. $2\dfrac{7}{6} \times 2\dfrac{5}{7} =$

17. $\dfrac{7}{6} \times 2\dfrac{5}{8} =$ 18. $1\dfrac{3}{4} \times 2\dfrac{10}{9} =$

Multiply fractions and write the answers in simplest form.

1. $2\dfrac{1}{5} \times \dfrac{7}{3} =$

2. $\dfrac{3}{4} \times 1\dfrac{5}{4} =$

3. $\dfrac{1}{4} \times 2\dfrac{5}{6} =$

4. $1\dfrac{2}{3} \times 1\dfrac{5}{8} =$

5. $1\dfrac{2}{3} \times 1\dfrac{8}{7} =$

6. $1\dfrac{1}{2} \times \dfrac{5}{9} =$

7. $1\dfrac{1}{5} \times \dfrac{3}{5} =$

8. $2\dfrac{4}{3} \times 1\dfrac{2}{5} =$

9. $\dfrac{5}{7} \times 1\dfrac{5}{8} =$

10. $2\dfrac{3}{4} \times \dfrac{5}{7} =$

11. $1\dfrac{3}{2} \times 1\dfrac{2}{9} =$

12. $\dfrac{5}{8} \times 2\dfrac{3}{4} =$

13. $\dfrac{2}{3} \times 2\dfrac{1}{3} =$

14. $1\dfrac{7}{10} \times 1\dfrac{3}{8} =$

15. $2\dfrac{4}{5} \times \dfrac{4}{7} =$

16. $2\dfrac{4}{7} \times 2\dfrac{7}{8} =$

17. $2\dfrac{2}{9} \times 2\dfrac{5}{6} =$

18. $2\dfrac{3}{5} \times 1\dfrac{1}{4} =$

Multiply fractions and write the answers in simplest form.

1. $1\dfrac{2}{5} \times 2\dfrac{3}{5} =$

2. $2\dfrac{1}{6} \times 1\dfrac{2}{5} =$

3. $2\dfrac{1}{3} \times 2\dfrac{1}{6} =$

4. $3\dfrac{3}{4} \times 1\dfrac{4}{5} =$

5. $1\dfrac{3}{4} \times 3\dfrac{2}{9} =$

6. $1\dfrac{1}{2} \times 3\dfrac{5}{8} =$

7. $2\dfrac{1}{6} \times 3\dfrac{1}{6} =$

8. $3\dfrac{4}{3} \times 3\dfrac{5}{4} =$

9. $3\dfrac{4}{3} \times 1\dfrac{3}{4} =$

10. $2\dfrac{1}{4} \times 3\dfrac{7}{9} =$

11. $2\dfrac{4}{5} \times 2\dfrac{5}{6} =$

12. $2\dfrac{3}{5} \times 2\dfrac{9}{10} =$

13. $1\dfrac{1}{2} \times 1\dfrac{5}{6} =$

14. $1\dfrac{7}{6} \times 2\dfrac{8}{7} =$

15. $3\dfrac{1}{2} \times 3\dfrac{4}{5} =$

16. $3\dfrac{3}{4} \times 3\dfrac{5}{6} =$

17. $2\dfrac{2}{3} \times 1\dfrac{4}{7} =$

18. $1\dfrac{11}{10} \times 2\dfrac{5}{6} =$

Multiply fractions and write the answers in simplest form.

1. $2\dfrac{1}{3} \times 2\dfrac{1}{4} =$ 2. $2\dfrac{1}{2} \times 2\dfrac{1}{4} =$

3. $3\dfrac{2}{3} \times 1\dfrac{1}{5} =$ 4. $3\dfrac{2}{3} \times 1\dfrac{1}{3} =$

5. $1\dfrac{3}{4} \times 2\dfrac{2}{5} =$ 6. $1\dfrac{3}{4} \times 2\dfrac{5}{6} =$

7. $2\dfrac{1}{4} \times 2\dfrac{5}{6} =$ 8. $2\dfrac{1}{6} \times 3\dfrac{7}{12} =$

9. $3\dfrac{3}{5} \times 2\dfrac{3}{10} =$ 10. $3\dfrac{2}{5} \times 3\dfrac{1}{6} =$

11. $3\dfrac{2}{5} \times 3\dfrac{5}{6} =$ 12. $3\dfrac{5}{6} \times 2\dfrac{4}{9} =$

13. $1\dfrac{3}{4} \times 3\dfrac{9}{10} =$ 14. $1\dfrac{3}{7} \times 1\dfrac{4}{5} =$

15. $2\dfrac{2}{3} \times 2\dfrac{7}{9} =$ 16. $2\dfrac{1}{6} \times 1\dfrac{6}{7} =$

17. $3\dfrac{3}{4} \times 1\dfrac{9}{10} =$ 18. $3\dfrac{5}{6} \times 3\dfrac{7}{9} =$

Multiply fractions and write the answers in simplest form.

1. $4\frac{1}{2} \times 3\frac{2}{3} =$

2. $3\frac{1}{2} \times 3\frac{1}{2} =$

3. $3\frac{1}{3} \times 4\frac{2}{5} =$

4. $4\frac{2}{3} \times 2\frac{1}{2} =$

5. $3\frac{2}{3} \times 2\frac{3}{4} =$

6. $1\frac{3}{4} \times 3\frac{2}{3} =$

7. $4\frac{1}{6} \times 2\frac{2}{3} =$

8. $2\frac{1}{5} \times 2\frac{2}{5} =$

9. $3\frac{3}{5} \times 4\frac{1}{2} =$

10. $2\frac{3}{4} \times 4\frac{3}{5} =$

11. $4\frac{1}{5} \times 1\frac{3}{4} =$

12. $3\frac{1}{3} \times 1\frac{1}{6} =$

13. $2\frac{2}{3} \times 3\frac{1}{3} =$

14. $1\frac{3}{5} \times 3\frac{1}{3} =$

15. $4\frac{2}{5} \times 2\frac{3}{5} =$

16. $2\frac{5}{6} \times 2\frac{3}{4} =$

17. $3\frac{1}{4} \times 3\frac{2}{5} =$

18. $4\frac{1}{3} \times 3\frac{4}{5} =$

Multiply fractions and write the answers in simplest form.

1. $4 \dfrac{1}{2} \times 2 \dfrac{2}{3} =$

2. $5 \dfrac{1}{3} \times 1 \dfrac{3}{2} =$

3. $3 \dfrac{3}{4} \times 4 \dfrac{1}{5} =$

4. $3 \dfrac{3}{4} \times 3 \dfrac{2}{3} =$

5. $5 \dfrac{2}{3} \times 1 \dfrac{2}{3} =$

6. $2 \dfrac{4}{5} \times 5 \dfrac{1}{2} =$

7. $2 \dfrac{2}{5} \times 3 \dfrac{1}{6} =$

8. $2 \dfrac{1}{2} \times 3 \dfrac{7}{6} =$

9. $3 \dfrac{2}{3} \times 1 \dfrac{8}{9} =$

10. $1 \dfrac{3}{4} \times 5 \dfrac{4}{3} =$

11. $2 \dfrac{3}{4} \times 5 \dfrac{3}{8} =$

12. $5 \dfrac{2}{3} \times 2 \dfrac{7}{8} =$

13. $5 \dfrac{1}{3} \times 3 \dfrac{2}{5} =$

14. $4 \dfrac{1}{4} \times 3 \dfrac{2}{5} =$

15. $4 \dfrac{1}{3} \times 3 \dfrac{3}{7} =$

16. $3 \dfrac{2}{5} \times 2 \dfrac{5}{6} =$

17. $2 \dfrac{5}{6} \times 2 \dfrac{5}{8} =$

18. $4 \dfrac{3}{4} \times 4 \dfrac{2}{3} =$

Multiply fractions and write the answers in simplest form.

1. $6\frac{1}{2} \times 7\frac{1}{2} =$

2. $6\frac{1}{2} \times 2\frac{1}{3} =$

3. $3\frac{2}{3} \times 5\frac{1}{3} =$

4. $5\frac{2}{3} \times 5\frac{1}{3} =$

5. $5\frac{2}{3} \times 6\frac{1}{4} =$

6. $4\frac{1}{4} \times 4\frac{2}{3} =$

7. $5\frac{1}{3} \times 3\frac{2}{3} =$

8. $5\frac{1}{2} \times 3\frac{3}{4} =$

9. $4\frac{1}{2} \times 6\frac{1}{3} =$

10. $2\frac{4}{5} \times 6\frac{1}{3} =$

11. $6\frac{3}{4} \times 6\frac{1}{2} =$

12. $5\frac{1}{3} \times 4\frac{3}{4} =$

13. $7\frac{1}{2} \times 6\frac{1}{3} =$

14. $3\frac{4}{5} \times 5\frac{2}{3} =$

15. $6\frac{2}{3} \times 5\frac{1}{2} =$

16. $4\frac{2}{3} \times 3\frac{2}{3} =$

17. $6\frac{1}{4} \times 4\frac{3}{4} =$

18. $6\frac{1}{2} \times 6\frac{3}{4} =$

Multiply fractions and write the answers in simplest form.

1. $7\dfrac{1}{2} \times 4\dfrac{1}{3} =$

2. $8\dfrac{1}{2} \times 5\dfrac{1}{2} =$

3. $8\dfrac{1}{3} \times 5\dfrac{2}{3} =$

4. $5\dfrac{1}{3} \times 7\dfrac{1}{2} =$

5. $4\dfrac{1}{2} \times 6\dfrac{1}{2} =$

6. $3\dfrac{2}{3} \times 5\dfrac{1}{3} =$

7. $6\dfrac{2}{3} \times 5\dfrac{1}{3} =$

8. $4\dfrac{1}{2} \times 6\dfrac{1}{2} =$

9. $6\dfrac{1}{2} \times 2\dfrac{1}{2} =$

10. $7\dfrac{2}{3} \times 5\dfrac{1}{3} =$

11. $5\dfrac{2}{3} \times 6\dfrac{1}{2} =$

12. $5\dfrac{1}{2} \times 5\dfrac{1}{2} =$

13. $6\dfrac{1}{2} \times 6\dfrac{2}{3} =$

14. $6\dfrac{2}{3} \times 3\dfrac{1}{2} =$

15. $7\dfrac{1}{3} \times 3\dfrac{2}{3} =$

16. $8\dfrac{1}{3} \times 9\dfrac{2}{3} =$

17. $7\dfrac{1}{2} \times 5\dfrac{1}{2} =$

18. $7\dfrac{2}{3} \times 7\dfrac{1}{2} =$

Divide fractions and write the answers in simplest form.

1. $2 \div \dfrac{1}{3} =$

2. $\dfrac{1}{2} \div 2 =$

3. $2 \div \dfrac{3}{2} =$

4. $\dfrac{1}{2} \div 3 =$

5. $2 \div \dfrac{1}{4} =$

6. $\dfrac{1}{2} \div 4 =$

7. $2 \div \dfrac{2}{5} =$

8. $\dfrac{1}{3} \div 2 =$

9. $3 \div \dfrac{3}{5} =$

10. $\dfrac{2}{3} \div 3 =$

11. $3 \div \dfrac{3}{4} =$

12. $\dfrac{1}{4} \div 2 =$

13. $3 \div \dfrac{4}{5} =$

14. $\dfrac{3}{4} \div 3 =$

15. $4 \div \dfrac{1}{6} =$

16. $\dfrac{1}{5} \div 2 =$

17. $4 \div \dfrac{5}{6} =$

18. $\dfrac{2}{5} \div 3 =$

Divide fractions and write the answers in simplest form.

1. $2 \div 2\dfrac{1}{2} =$ 2. $1\dfrac{1}{2} \div 2 =$

3. $3 \div 2\dfrac{1}{3} =$ 4. $1\dfrac{1}{3} \div 2 =$

5. $2 \div 1\dfrac{2}{3} =$ 6. $2\dfrac{2}{3} \div 2 =$

7. $1 \div 1\dfrac{1}{4} =$ 8. $2\dfrac{1}{5} \div 3 =$

9. $3 \div 1\dfrac{3}{4} =$ 10. $1\dfrac{1}{4} \div 3 =$

11. $1 \div 2\dfrac{1}{5} =$ 12. $2\dfrac{3}{4} \div 3 =$

13. $2 \div 2\dfrac{2}{5} =$ 14. $1\dfrac{2}{5} \div 2 =$

15. $3 \div 1\dfrac{3}{5} =$ 16. $1\dfrac{1}{6} \div 3 =$

17. $2 \div 1\dfrac{4}{5} =$ 18. $2\dfrac{3}{5} \div 3 =$

Divide fractions and write the answers in simplest form.

1. $\dfrac{1}{2} \div 1\dfrac{1}{2} =$

2. $1\dfrac{1}{2} \div \dfrac{2}{3} =$

3. $\dfrac{3}{4} \div \dfrac{5}{6} =$

4. $\dfrac{1}{4} \div 1\dfrac{3}{4} =$

5. $\dfrac{2}{3} \div 1\dfrac{1}{5} =$

6. $1\dfrac{1}{3} \div \dfrac{2}{5} =$

7. $\dfrac{2}{3} \div \dfrac{3}{4} =$

8. $1\dfrac{2}{3} \div \dfrac{3}{5} =$

9. $\dfrac{2}{3} \div 1\dfrac{1}{3} =$

10. $\dfrac{2}{5} \div 1\dfrac{1}{2} =$

11. $\dfrac{4}{5} \div \dfrac{5}{6} =$

12. $1\dfrac{1}{5} \div 2\dfrac{1}{3} =$

13. $\dfrac{5}{6} \div 1\dfrac{1}{11} =$

14. $\dfrac{3}{4} \div 1\dfrac{2}{3} =$

15. $\dfrac{4}{5} \div 1\dfrac{1}{5} =$

16. $2\dfrac{3}{7} \div \dfrac{3}{4} =$

17. $\dfrac{7}{9} \div 1\dfrac{2}{5} =$

18. $\dfrac{4}{5} \div 1\dfrac{1}{3} =$

101

Name: _____   Lesson 10-4   Dividing fractions

Divide fractions and write the answers in simplest form.

1. $\dfrac{1}{2} \div \dfrac{7}{6} =$

2. $\dfrac{4}{5} \div \dfrac{8}{5} =$

3. $\dfrac{3}{4} \div \dfrac{10}{9} =$

4. $\dfrac{9}{4} \div \dfrac{5}{8} =$

5. $\dfrac{11}{12} \div \dfrac{15}{14} =$

6. $\dfrac{9}{10} \div \dfrac{9}{20} =$

7. $\dfrac{2}{3} \div \dfrac{7}{6} =$

8. $\dfrac{5}{6} \div \dfrac{9}{10} =$

9. $\dfrac{5}{6} \div \dfrac{9}{8} =$

10. $\dfrac{8}{9} \div \dfrac{18}{17} =$

11. $\dfrac{11}{12} \div \dfrac{11}{10} =$

12. $\dfrac{7}{8} \div \dfrac{12}{11} =$

13. $\dfrac{3}{4} \div \dfrac{7}{8} =$

14. $\dfrac{4}{5} \div \dfrac{8}{7} =$

15. $\dfrac{5}{8} \div \dfrac{9}{16} =$

16. $\dfrac{5}{6} \div \dfrac{15}{14} =$

17. $\dfrac{11}{15} \div \dfrac{7}{30} =$

18. $\dfrac{5}{8} \div \dfrac{16}{5} =$

Name:

Lesson 10-5   Dividing fractions

Divide fractions and write the answers in simplest form.

1. $\dfrac{1}{2} \div \dfrac{2}{3} =$

2. $\dfrac{5}{3} \div \dfrac{15}{7} =$

3. $\dfrac{2}{3} \div \dfrac{5}{6} =$

4. $\dfrac{3}{2} \div \dfrac{5}{4} =$

5. $\dfrac{1}{4} \div \dfrac{12}{7} =$

6. $\dfrac{4}{3} \div \dfrac{3}{4} =$

7. $\dfrac{1}{3} \div \dfrac{9}{5} =$

8. $\dfrac{5}{2} \div \dfrac{8}{3} =$

9. $\dfrac{3}{4} \div \dfrac{8}{9} =$

10. $\dfrac{7}{4} \div \dfrac{9}{14} =$

11. $\dfrac{5}{6} \div \dfrac{12}{5} =$

12. $\dfrac{4}{5} \div \dfrac{3}{10} =$

13. $\dfrac{2}{5} \div \dfrac{15}{4} =$

14. $\dfrac{9}{4} \div \dfrac{7}{2} =$

15. $\dfrac{3}{7} \div \dfrac{14}{5} =$

16. $\dfrac{10}{3} \div \dfrac{6}{5} =$

17. $\dfrac{4}{5} \div \dfrac{8}{3} =$

18. $\dfrac{11}{7} \div \dfrac{5}{14} =$

103

Divide fractions and write the answers in simplest form.

1. $1\dfrac{1}{3} \div \dfrac{2}{3} =$

2. $2\dfrac{1}{3} \div 1\dfrac{1}{2} =$

3. $1\dfrac{1}{2} \div 1\dfrac{1}{3} =$

4. $1\dfrac{3}{4} \div \dfrac{2}{3} =$

5. $\dfrac{3}{4} \div 1\dfrac{2}{3} =$

6. $\dfrac{1}{2} \div 2\dfrac{1}{4} =$

7. $1\dfrac{1}{5} \div 2\dfrac{1}{2} =$

8. $2\dfrac{3}{5} \div \dfrac{5}{2} =$

9. $2\dfrac{1}{4} \div 1\dfrac{2}{3} =$

10. $\dfrac{2}{5} \div \dfrac{3}{4} =$

11. $1\dfrac{2}{5} \div 1\dfrac{1}{4} =$

12. $1\dfrac{1}{2} \div 1\dfrac{2}{5} =$

13. $\dfrac{3}{7} \div 1\dfrac{1}{6} =$

14. $2\dfrac{4}{5} \div 2\dfrac{2}{5} =$

15. $2\dfrac{3}{5} \div 2\dfrac{1}{3} =$

16. $2\dfrac{1}{3} \div \dfrac{5}{6} =$

17. $1\dfrac{5}{6} \div \dfrac{11}{7} =$

18. $\dfrac{3}{4} \div 1\dfrac{1}{2} =$

Divide fractions and write the answers in simplest form.

1. $2\dfrac{1}{2} \div 2\dfrac{1}{4} =$ 2. $1\dfrac{3}{5} \div 2\dfrac{1}{3} =$

3. $3\dfrac{2}{3} \div \dfrac{2}{3} =$ 4. $1\dfrac{2}{3} \div \dfrac{2}{5} =$

5. $1\dfrac{3}{5} \div 3\dfrac{1}{6} =$ 6. $3\dfrac{1}{4} \div 2\dfrac{1}{3} =$

7. $2\dfrac{1}{3} \div 2\dfrac{2}{5} =$ 8. $1\dfrac{2}{5} \div 1\dfrac{3}{4} =$

9. $3\dfrac{1}{2} \div 2\dfrac{3}{4} =$ 10. $3\dfrac{1}{2} \div 1\dfrac{2}{3} =$

11. $\dfrac{1}{4} \div \dfrac{2}{5} =$ 12. $2\dfrac{3}{4} \div 3\dfrac{1}{4} =$

13. $\dfrac{2}{7} \div 2\dfrac{1}{3} =$ 14. $3\dfrac{4}{5} \div \dfrac{2}{5} =$

15. $2\dfrac{4}{5} \div 3\dfrac{1}{2} =$ 16. $3\dfrac{3}{7} \div 3\dfrac{1}{2} =$

17. $3\dfrac{3}{7} \div 3\dfrac{2}{3} =$ 18. $\dfrac{5}{6} \div 3\dfrac{1}{2} =$

Divide fractions and write the answers in simplest form.

1. $3\frac{1}{2} \div \frac{2}{3} =$ 2. $3\frac{3}{4} \div 1\frac{1}{3} =$

3. $4\frac{1}{3} \div 2\frac{1}{2} =$ 4. $2\frac{2}{5} \div 2\frac{1}{5} =$

5. $5\frac{2}{3} \div 2\frac{1}{3} =$ 6. $4\frac{2}{3} \div 3\frac{1}{2} =$

7. $3\frac{1}{4} \div 2\frac{3}{4} =$ 8. $5\frac{1}{6} \div 2\frac{1}{3} =$

9. $5\frac{3}{5} \div 3\frac{1}{3} =$ 10. $3\frac{4}{5} \div 2\frac{2}{3} =$

11. $2\frac{1}{6} \div 1\frac{2}{3} =$ 12. $5\frac{3}{7} \div 3\frac{2}{5} =$

13. $4\frac{2}{5} \div 3\frac{1}{4} =$ 14. $4\frac{1}{4} \div 1\frac{1}{2} =$

15. $4\frac{3}{4} \div 1\frac{2}{3} =$ 16. $3\frac{2}{3} \div 3\frac{1}{4} =$

17. $5\frac{5}{6} \div 4\frac{1}{6} =$ 18. $5\frac{3}{4} \div 3\frac{2}{5} =$

Divide fractions and write the answers in simplest form.

1. $3\frac{1}{2} \div 4\frac{2}{3} =$

2. $2\frac{1}{2} \div 3\frac{3}{4} =$

3. $5\frac{1}{3} \div 2\frac{2}{3} =$

4. $1\frac{2}{3} \div 4\frac{1}{2} =$

5. $2\frac{2}{3} \div 3\frac{1}{2} =$

6. $2\frac{1}{4} \div 4\frac{1}{3} =$

7. $2\frac{1}{5} \div 6\frac{1}{2} =$

8. $1\frac{1}{3} \div 3\frac{1}{2} =$

9. $1\frac{3}{5} \div 2\frac{2}{3} =$

10. $3\frac{3}{4} \div 4\frac{1}{3} =$

11. $2\frac{1}{2} \div 4\frac{1}{5} =$

12. $4\frac{2}{3} \div 5\frac{2}{5} =$

13. $3\frac{2}{3} \div 4\frac{1}{2} =$

14. $2\frac{1}{4} \div 4\frac{1}{5} =$

15. $2\frac{1}{3} \div 3\frac{2}{3} =$

16. $4\frac{2}{5} \div 5\frac{1}{3} =$

17. $4\frac{3}{5} \div 5\frac{4}{5} =$

18. $3\frac{3}{4} \div 6\frac{1}{2} =$

Divide fractions and write the answers in simplest form.

1. $7\dfrac{1}{2} \div 4\dfrac{1}{2} =$ 2. $2\dfrac{1}{3} \div 5\dfrac{1}{2} =$

3. $5\dfrac{1}{2} \div 2\dfrac{1}{3} =$ 4. $3\dfrac{1}{2} \div 5\dfrac{1}{3} =$

5. $6\dfrac{2}{3} \div 5\dfrac{1}{2} =$ 6. $4\dfrac{2}{3} \div 7\dfrac{1}{3} =$

7. $5\dfrac{2}{3} \div 2\dfrac{2}{3} =$ 8. $2\dfrac{1}{2} \div 5\dfrac{1}{4} =$

9. $5\dfrac{1}{5} \div 3\dfrac{1}{2} =$ 10. $6\dfrac{2}{3} \div 8\dfrac{3}{4} =$

11. $8\dfrac{1}{4} \div 6\dfrac{2}{3} =$ 12. $4\dfrac{3}{4} \div 7\dfrac{1}{2} =$

13. $7\dfrac{1}{3} \div 3\dfrac{1}{2} =$ 14. $5\dfrac{1}{4} \div 6\dfrac{1}{3} =$

15. $9\dfrac{3}{5} \div 5\dfrac{1}{3} =$ 16. $2\dfrac{1}{3} \div 6\dfrac{1}{2} =$

17. $8\dfrac{2}{5} \div 2\dfrac{2}{3} =$ 18. $5\dfrac{3}{4} \div 9\dfrac{3}{4} =$

Name: _____   Lesson 11-1   Ratios

Change the following ratios to fractions and write the answer in simplest form.

Example: $2:3$ or 2 to 3 $= \dfrac{2}{3}$

1. $1:3 =$ _____

2. $2:7 =$ _____

3. $3:13 =$ _____

4. $2:4 =$ _____

5. $3:4 =$ _____

6. $6:12 =$ _____

7. $1:5 =$ _____

8. $1:10 =$ _____

9. $4:13 =$ _____

10. $2:5 =$ _____

11. $4:10 =$ _____

12. $5:12 =$ _____

13. $3:7 =$ _____

14. $3:9 =$ _____

15. $3:15 =$ _____

16. $4:6 =$ _____

17. $5:9 =$ _____

18. $2:10 =$ _____

19. $3:6 =$ _____

20. $6:10 =$ _____

21. $4:16 =$ _____

22. $3:8 =$ _____

23. $2:11 =$ _____

24. $7:14 =$ _____

25. $3:9 =$ _____

26. $4:12 =$ _____

27. $8:24 =$ _____

Change the following ratios to fractions and write the answer in simplest form.

Example: $2:3$ or 2 to 3 $= \dfrac{2}{3}$

1. $3:4 = $ _____

2. $2:3 = $ _____

3. $3:9 = $ _____

4. $2:6 = $ _____

5. $4:5 = $ _____

6. $3:7 = $ _____

7. $1:3 = $ _____

8. $3:6 = $ _____

9. $2:10 = $ _____

10. $3:5 = $ _____

11. $4:7 = $ _____

12. $4:6 = $ _____

13. $2:4 = $ _____

14. $2:8 = $ _____

15. $3:8 = $ _____

16. $2:7 = $ _____

17. $5:10 = $ _____

18. $5:15 = $ _____

19. $4:8 = $ _____

20. $6:18 = $ _____

21. $2:12 = $ _____

22. $5:9 = $ _____

23. $2:9 = $ _____

24. $6:10 = $ _____

25. $8:16 = $ _____

26. $7:12 = $ _____

27. $7:21 = $ _____

Change the following fractions to ratios.

Example: $\dfrac{2}{3} = 2:3$ or 2 to 3

1. $\dfrac{1}{2}$ = _____

2. $\dfrac{1}{3}$ = _____

3. $\dfrac{1}{4}$ = _____

4. $\dfrac{2}{3}$ = _____

5. $\dfrac{4}{3}$ = _____

6. $\dfrac{5}{3}$ = _____

7. $\dfrac{3}{2}$ = _____

8. $\dfrac{3}{4}$ = _____

9. $\dfrac{3}{5}$ = _____

10. $\dfrac{1}{4}$ = _____

11. $\dfrac{5}{2}$ = _____

12. $\dfrac{7}{10}$ = _____

13. $\dfrac{2}{5}$ = _____

14. $\dfrac{4}{7}$ = _____

15. $\dfrac{4}{9}$ = _____

16. $\dfrac{5}{7}$ = _____

17. $\dfrac{4}{5}$ = _____

18. $\dfrac{10}{9}$ = _____

19. $\dfrac{2}{9}$ = _____

20. $\dfrac{7}{6}$ = _____

21. $\dfrac{5}{4}$ = _____

22. $\dfrac{10}{3}$ = _____

23. $\dfrac{4}{9}$ = _____

24. $\dfrac{4}{11}$ = _____

25. $\dfrac{7}{12}$ = _____

26. $\dfrac{3}{10}$ = _____

27. $\dfrac{9}{4}$ = _____

Change the following fractions to ratios.

Example: $\dfrac{2}{3} = 2 : 3$ or 2 to 3

1. $\dfrac{4}{5} =$ _____

2. $\dfrac{1}{5} =$ _____

3. $\dfrac{5}{6} =$ _____

4. $\dfrac{2}{5} =$ _____

5. $\dfrac{5}{4} =$ _____

6. $\dfrac{7}{8} =$ _____

7. $\dfrac{3}{4} =$ _____

8. $\dfrac{9}{10} =$ _____

9. $\dfrac{11}{12} =$ _____

10. $\dfrac{8}{3} =$ _____

11. $\dfrac{11}{7} =$ _____

12. $\dfrac{6}{5} =$ _____

13. $\dfrac{2}{7} =$ _____

14. $\dfrac{7}{2} =$ _____

15. $\dfrac{12}{11} =$ _____

16. $\dfrac{5}{2} =$ _____

17. $\dfrac{3}{8} =$ _____

18. $\dfrac{8}{7} =$ _____

19. $\dfrac{10}{9} =$ _____

20. $\dfrac{11}{9} =$ _____

21. $\dfrac{3}{14} =$ _____

22. $\dfrac{2}{13} =$ _____

23. $\dfrac{15}{8} =$ _____

24. $\dfrac{8}{13} =$ _____

25. $\dfrac{14}{5} =$ _____

26. $\dfrac{9}{16} =$ _____

27. $\dfrac{15}{7} =$ _____

Reduce the following ratios to their lowest forms.

Example: $4 : 6 = 2 : 3$

1. $6 : 4 =$ _____

2. $2 : 4 =$ _____

3. $6 : 3 =$ _____

4. $3 : 6 =$ _____

5. $4 : 12 =$ _____

6. $5 : 10 =$ _____

7. $4 : 8 =$ _____

8. $6 : 9 =$ _____

9. $4 : 16 =$ _____

10. $10 : 15 =$ _____

11. $8 : 12 =$ _____

12. $6 : 8 =$ _____

13. $8 : 16 =$ _____

14. $12 : 15 =$ _____

15. $9 : 18 =$ _____

16. $18 : 27 =$ _____

17. $16 : 24 =$ _____

18. $14 : 21 =$ _____

19. $12 : 16 =$ _____

20. $15 : 9 =$ _____

21. $20 : 12 =$ _____

22. $20 : 15 =$ _____

23. $12 : 18 =$ _____

24. $35 : 21 =$ _____

25. $30 : 12 =$ _____

26. $21 : 18 =$ _____

27. $24 : 28 =$ _____

Reduce the following ratios to their lowest forms.

Example: $4 : 6 = 2 : 3$

1. $12 : 3 =$ _____

2. $6 : 10 =$ _____

3. $12 : 10 =$ _____

4. $6 : 18 =$ _____

5. $42 : 7 =$ _____

6. $12 : 32 =$ _____

7. $21 : 14 =$ _____

8. $16 : 48 =$ _____

9. $40 : 20 =$ _____

10. $32 : 8 =$ _____

11. $24 : 30 =$ _____

12. $32 : 12 =$ _____

13. $18 : 36 =$ _____

14. $12 : 8 =$ _____

15. $9 : 81 =$ _____

16. $6 : 8 =$ _____

17. $18 : 15 =$ _____

18. $42 : 56 =$ _____

19. $3 : 15 =$ _____

20. $15 : 35 =$ _____

21. $13 : 39 =$ _____

22. $20 : 8 =$ _____

23. $18 : 42 =$ _____

24. $18 : 4 =$ _____

25. $15 : 25 =$ _____

26. $49 : 28 =$ _____

27. $22 : 33 =$ _____

Reduce the following ratios to their lowest forms.

Example: $4 : 6 = 2 : 3$

1. $12 : 6 =$ _____

2. $7 : 21 =$ _____

3. $24 : 16 =$ _____

4. $9 : 36 =$ _____

5. $4 : 16 =$ _____

6. $9 : 12 =$ _____

7. $3 : 15 =$ _____

8. $25 : 10 =$ _____

9. $18 : 30 =$ _____

10. $54 : 18 =$ _____

11. $28 : 35 =$ _____

12. $42 : 12 =$ _____

13. $20 : 24 =$ _____

14. $12 : 8 =$ _____

15. $45 : 63 =$ _____

16. $21 : 49 =$ _____

17. $5 : 45 =$ _____

18. $35 : 65 =$ _____

19. $40 : 15 =$ _____

20. $32 : 56 =$ _____

21. $24 : 36 =$ _____

22. $64 : 48 =$ _____

23. $48 : 32 =$ _____

24. $52 : 39 =$ _____

25. $6 : 27 =$ _____

26. $72 : 63 =$ _____

27. $33 : 55 =$ _____

Find equivalent ratios.

Example: $1 : 2 = 2 : 4 = 3 : 6 = 4 : 8$

1. $2 : 3 = 4 : \underline{} = 6 : \underline{} = 8 : \underline{} = 12 : \underline{} = 24 : \underline{}$

2. $1 : 3 = \underline{} : 6 = \underline{} : 9 = \underline{} : 15 = \underline{} : 18 = \underline{} : 27$

3. $3 : 2 = 9 : \underline{} = 15 : \underline{} = \underline{} : 12 = \underline{} : 16 = 30 : \underline{}$

4. $1 : 4 = 2 : \underline{} = \underline{} : 12 = 5 : \underline{} = \underline{} : 24 = \underline{} : 32$

5. $4 : 3 = 12 : \underline{} = 16 : \underline{} = \underline{} : 18 = \underline{} : 21 = 36 : \underline{}$

6. $2 : 5 = \underline{} : 20 = \underline{} : 25 = 14 : \underline{} = 16 : \underline{} = 22 : \underline{}$

7. $3 : 4 = 9 : \underline{} = 15 : \underline{} = \underline{} : 24 = \underline{} : 32 = 36 : \underline{}$

8. $1 : 5 = \underline{} : 10 = \underline{} : 15 = \underline{} : 25 = \underline{} : 30 = 8 : \underline{}$

9. $5 : 2 = 10 : \underline{} = \underline{} : 8 = 25 : \underline{} = \underline{} : 12 = 45 : \underline{}$

10. $2 : 7 = 4 : \underline{} = 8 : \underline{} = \underline{} : 35 = \underline{} : 49 = 16 : \underline{}$

11. $8 : 3 = 24 : \underline{} = 32 : \underline{} = \underline{} : 18 = \underline{} : 21 = 72 : \underline{}$

12. $3 : 5 = 6 : \underline{} = \underline{} : 15 = 12 : \underline{} = \underline{} : 30 = \underline{} : 45$

13. $5 : 6 = \underline{} : 18 = 20 : \underline{} = 25 : \underline{} = \underline{} : 42 = \underline{} : 48$

Find equivalent ratios.

Example: $1 : 2 = 2 : 4 = 3 : 6 = 4 : 8$

1. $2 : 3 = 4 : \rule{1cm}{0.4pt} = 6 : \rule{1cm}{0.4pt} = 8 : \rule{1cm}{0.4pt} = 12 : \rule{1cm}{0.4pt} = 16 : \rule{1cm}{0.4pt}$

2. $2 : 5 = \rule{1cm}{0.4pt} : 15 = 10 : \rule{1cm}{0.4pt} = \rule{1cm}{0.4pt} : 35 = 16 : \rule{1cm}{0.4pt} = 26 : \rule{1cm}{0.4pt}$

3. $4 : 1 = 8 : \rule{1cm}{0.4pt} = \rule{1cm}{0.4pt} : 4 = 20 : \rule{1cm}{0.4pt} = \rule{1cm}{0.4pt} : 7 = 36 : \rule{1cm}{0.4pt}$

4. $3 : 2 = \rule{1cm}{0.4pt} : 6 = 12 : \rule{1cm}{0.4pt} = \rule{1cm}{0.4pt} : 10 = 21 : \rule{1cm}{0.4pt} = \rule{1cm}{0.4pt} : 18$

5. $10 : 4 = 5 : \rule{1cm}{0.4pt} = 15 : \rule{1cm}{0.4pt} = \rule{1cm}{0.4pt} : 16 = 30 : \rule{1cm}{0.4pt} = \rule{1cm}{0.4pt} : 32$

6. $3 : 7 = 9 : \rule{1cm}{0.4pt} = \rule{1cm}{0.4pt} : 28 = \rule{1cm}{0.4pt} : 35 = 18 : \rule{1cm}{0.4pt} = \rule{1cm}{0.4pt} : 77$

7. $5 : 4 = \rule{1cm}{0.4pt} : 12 = \rule{1cm}{0.4pt} : 16 = \rule{1cm}{0.4pt} : 20 = 30 : \rule{1cm}{0.4pt} = 45 : \rule{1cm}{0.4pt}$

8. $4 : 3 = 8 : \rule{1cm}{0.4pt} = 12 : \rule{1cm}{0.4pt} = \rule{1cm}{0.4pt} : 12 = \rule{1cm}{0.4pt} : 18 = 28 : \rule{1cm}{0.4pt}$

9. $4 : 6 = 2 : \rule{1cm}{0.4pt} = 6 : \rule{1cm}{0.4pt} = \rule{1cm}{0.4pt} : 12 = 12 : \rule{1cm}{0.4pt} = \rule{1cm}{0.4pt} : 24$

10. $7 : 4 = 28 : \rule{1cm}{0.4pt} = \rule{1cm}{0.4pt} : 20 = 42 : \rule{1cm}{0.4pt} = 56 : \rule{1cm}{0.4pt} = \rule{1cm}{0.4pt} : 36$

11. $8 : 4 = \rule{1cm}{0.4pt} : 2 = 2 : \rule{1cm}{0.4pt} = 10 : \rule{1cm}{0.4pt} = \rule{1cm}{0.4pt} : 6 = 16 : \rule{1cm}{0.4pt}$

12. $3 : 1 = 9 : \rule{1cm}{0.4pt} = 12 : \rule{1cm}{0.4pt} = \rule{1cm}{0.4pt} : 5 = \rule{1cm}{0.4pt} : 8 = 33 : \rule{1cm}{0.4pt}$

13. $12 : 2 = 6 : \rule{1cm}{0.4pt} = 18 : \rule{1cm}{0.4pt} = \rule{1cm}{0.4pt} : 5 = \rule{1cm}{0.4pt} : 12 = \rule{1cm}{0.4pt} : 16$

Find equivalent ratios.

Example: $1 : 2 = 2 : 4 = 3 : 6 = 4 : 8$

1. $2 : 1 = \underline{} : 2 = 10 : \underline{} = 12 : \underline{} = 18 : \underline{} = \underline{} : 15$

2. $3 : 2 = \underline{} : 4 = \underline{} : 6 = 15 : \underline{} = 27 : \underline{} = \underline{} : 26$

3. $10 : 4 = 5 : \underline{} = \underline{} : 10 = 35 : \underline{} = \underline{} : 18 = 70 : \underline{}$

4. $4 : 6 = \underline{} : 3 = 10 : \underline{} = 12 : \underline{} = \underline{} : 24 = \underline{} : 45$

5. $6 : 10 = 3 : \underline{} = 12 : \underline{} = \underline{} : 35 = 24 : \underline{} = \underline{} : 50$

6. $12 : 4 = \underline{} : 1 = 15 : \underline{} = 24 : \underline{} = \underline{} : 12 = 45 : \underline{}$

7. $15 : 12 = 5 : \underline{} = \underline{} : 8 = 30 : \underline{} = 45 : \underline{} = \underline{} : 48$

8. $9 : 21 = \underline{} : 7 = \underline{} : 28 = \underline{} : 56 = \underline{} : 63 = \underline{} : 84$

9. $16 : 12 = 4 : \underline{} = 8 : \underline{} = 28 : \underline{} = 40 : \underline{} = 56 : \underline{}$

10. $10 : 45 = \underline{} : 9 = 12 : \underline{} = 14 : \underline{} = \underline{} : 81 = \underline{} : 99$

11. $21 : 24 = 7 : \underline{} = 14 : \underline{} = \underline{} : 40 = \underline{} : 64 = 84 : \underline{}$

12. $36 : 45 = 4 : \underline{} = 12 : \underline{} = \underline{} : 25 = 28 : \underline{} = 60 : \underline{}$

13. $14 : 12 = \underline{} : 6 = 21 : \underline{} = \underline{} : 36 = 49 : \underline{} = 63 : \underline{}$

Evaluate the following exponents.

Example: $5^2 = 5 \times 5 = 25$, $4^3 = 4 \times 4 \times 4 = 64$, $2^4 = 2 \times 2 \times 2 \times 2 = 16$

1. $2^2 =$

2. $2^3 =$

3. $4^2 =$

4. $4^2 =$

5. $3^3 =$

6. $7^2 =$

7. $3^2 =$

8. $4^2 =$

9. $3^4 =$

10. $8^2 =$

11. $2^5 =$

12. $6^2 =$

13. $2^6 =$

14. $6^2 =$

15. $3^5 =$

16. $7^2 =$

17. $10^2 =$

18. $6^3 =$

19. $2^7 =$

20. $5^3 =$

21. $2^6 =$

22. $9^2 =$

23. $4^4 =$

24. $2^8 =$

25. $11^2 =$

26. $7^3 =$

27. $12^2 =$

Evaluate the following exponents.

Example: $5^2 = 5 \times 5 = 25$, $4^3 = 4 \times 4 \times 4 = 64$, $2^4 = 2 \times 2 \times 2 \times 2 = 16$

1. $5^2 =$

2. $3^2 =$

3. $4^2 =$

4. $3^3 =$

5. $7^2 =$

6. $2^6 =$

7. $8^2 =$

8. $9^2 =$

9. $5^3 =$

10. $6^3 =$

11. $2^7 =$

12. $2^9 =$

13. $2^{10} =$

14. $4^3 =$

15. $8^3 =$

16. $3^5 =$

17. $6^3 =$

18. $12^2 =$

19. $4^4 =$

20. $11^2 =$

21. $15^2 =$

22. $17^2 =$

23. $14^2 =$

24. $16^2 =$

25. $18^2 =$

26. $19^2 =$

27. $20^2 =$

Evaluate the following exponents.

Example: $5^2 = 5 \times 5 = 25$, $4^3 = 4 \times 4 \times 4 = 64$, $2^4 = 2 \times 2 \times 2 \times 2 = 16$

1. $5^2 =$ 2. $7^2 =$ 3. $4^2 =$

4. $3^3 =$ 5. $4^3 =$ 6. $5^3 =$

7. $2^5 =$ 8. $11^2 =$ 9. $9^2 =$

10. $2^9 =$ 11. $6^3 =$ 12. $12^2 =$

13. $6^2 =$ 14. $2^7 =$ 15. $10^2 =$

16. $14^2 =$ 17. $8^2 =$ 18. $7^3 =$

19. $15^2 =$ 20. $13^2 =$ 21. $2^8 =$

22. $21^2 =$ 23. $16^2 =$ 24. $10^3 =$

25. $9^3 =$ 26. $25^2 =$ 27. $22^2 =$

Evaluate the following exponents.

Example: $5^2 = 5 \times 5 = 25$, $4^3 = 4 \times 4 \times 4 = 64$, $2^4 = 2 \times 2 \times 2 \times 2 = 16$

1. $2^6 =$ 2. $4^3 =$ 3. $2^5 =$

4. $6^3 =$ 5. $3^5 =$ 6. $8^3 =$

7. $13^2 =$ 8. $2^{10} =$ 9. $10^2 =$

10. $14^2 =$ 11. $15^2 =$ 12. $16^2 =$

13. $2^9 =$ 14. $17^2 =$ 15. $24^2 =$

16. $20^2 =$ 17. $3^4 =$ 18. $7^2 =$

19. $30^2 =$ 20. $10^3 =$ 21. $18^2 =$

22. $9^3 =$ 23. $25^2 =$ 24. $23^2 =$

25. $12^2 =$ 26. $5^3 =$ 27. $17^2 =$

Evaluate the following exponents.

Example: $2^4 + 3^2 = 16 + 9 = 25$

1. $2^3 + 3^2 =$ 2. $2^2 + 5^2 =$ 3. $5^2 + 4^2 =$

4. $2^4 + 3^3 =$ 5. $4^2 + 6^2 =$ 6. $7^2 + 4^3 =$

7. $2^3 + 4^2 =$ 8. $3^3 + 6^2 =$ 9. $10^2 + 9^2 =$

10. $5^2 + 3^3 =$ 11. $4^3 + 2^3 =$ 12. $3^5 + 4^3 =$

13. $2^6 + 3^4 =$ 14. $2^5 + 3^4 =$ 15. $12^2 + 5^2 =$

16. $6^2 + 7^2 =$ 17. $3^4 + 5^3 =$ 18. $2^7 + 2^6 =$

19. $9^2 + 8^2 =$ 20. $12^2 + 2^3 =$ 21. $10^2 + 13^2 =$

22. $11^2 + 5^2 =$ 23. $8^2 + 12^2 =$ 24. $3^3 + 3^4 =$

25. $2^8 + 2^9 =$ 26. $5^3 + 6^3 =$ 27. $10^2 + 10^3 =$

Evaluate the following exponents.

Example: $2^4 + 3^2 = 16 + 9 = 25$

1. $3^4 + 4^2 =$

2. $8^2 + 9^2 =$

3. $3^2 + 4^2 =$

4. $12^2 + 5^2 =$

5. $5^2 + 10^2 =$

6. $9^2 + 11^2 =$

7. $8^2 + 4^3 =$

8. $2^4 + 2^5 =$

9. $6^2 + 8^2 =$

10. $3^3 + 5^3 =$

11. $12^2 + 14^2 =$

12. $2^8 + 14^2 =$

13. $2^6 + 9^2 =$

14. $5^3 + 10^2 =$

15. $15^2 + 20^2 =$

16. $8^2 + 15^2 =$

17. $6^3 + 3^5 =$

18. $6^2 + 7^2 =$

19. $13^2 + 19^2 =$

20. $11^2 + 12^2 =$

21. $10^2 + 20^2 =$

22. $2^5 + 2^7 =$

23. $2^9 + 2^8 =$

24. $12^2 + 17^2 =$

25. $7^3 + 16^2 =$

26. $5^3 + 9^3 =$

27. $8^3 + 16^2 =$

Evaluate the following exponents.

Example: $3^4 - 4^3 = 81 - 64 = 17$

1. $6^2 - 4^2 =$ 2. $9^2 - 8^2 =$ 3. $8^2 - 4^2 =$

4. $4^3 - 2^4 =$ 5. $11^2 - 3^4 =$ 6. $3^4 - 7^2 =$

7. $10^2 - 8^2 =$ 8. $6^3 - 4^3 =$ 9. $7^3 - 2^6 =$

10. $5^3 - 5^2 =$ 11. $14^2 - 13^2 =$ 12. $21^2 - 12^2 =$

13. $7^2 - 3^2 =$ 14. $2^8 - 5^3 =$ 15. $10^2 - 6^2 =$

16. $2^5 - 4^2 =$ 17. $15^2 - 13^2 =$ 18. $2^9 - 6^3 =$

19. $12^2 - 6^2 =$ 20. $3^5 - 15^2 =$ 21. $19^2 - 2^6 =$

22. $6^3 - 9^2 =$ 23. $14^2 - 3^4 =$ 24. $16^2 - 14^2 =$

25. $2^7 - 4^3 =$ 26. $20^2 - 17^2 =$ 27. $4^4 - 5^3 =$

Evaluate the following exponents.

Example: $3^4 - 4^3 = 81 - 64 = 17$

1. $9^2 - 3^2 =$

2. $10^2 - 4^2 =$

3. $7^2 - 3^2 =$

4. $3^4 - 6^2 =$

5. $4^3 - 2^5 =$

6. $12^2 - 9^2 =$

7. $5^3 - 2^6 =$

8. $12^2 - 7^2 =$

9. $3^4 - 4^2 =$

10. $10^2 - 5^2 =$

11. $7^3 - 15^2 =$

12. $15^2 - 11^2 =$

13. $2^8 - 4^3 =$

14. $2^8 - 4^3 =$

15. $2^8 - 5^3 =$

16. $5^3 - 4^3 =$

17. $10^3 - 2^9 =$

18. $7^3 - 2^7 =$

19. $11^2 - 9^2 =$

20. $8^3 - 6^3 =$

21. $3^5 - 5^3 =$

22. $6^3 - 13^2 =$

23. $4^4 - 3^5 =$

24. $4^4 - 15^2 =$

25. $15^2 - 2^7 =$

26. $6^3 - 5^3 =$

27. $9^3 - 25^2 =$

Evaluate the following exponents.

Example: $2^3 \times 3^2 = 8 \times 9 = 72$

1. $2^2 \times 3^2 =$ 2. $3^2 \times 3^3 =$ 3. $7^2 \times 2^2 =$

4. $3^2 \times 2^3 =$ 5. $4^2 \times 4^2 =$ 6. $8^2 \times 2^3 =$

7. $2^3 \times 4^2 =$ 8. $4^3 \times 2^2 =$ 9. $2^4 \times 3^3 =$

10. $3^2 \times 4^2 =$ 11. $2^5 \times 2^3 =$ 12. $10^2 \times 2^2 =$

13. $2^4 \times 2^2 =$ 14. $5^2 \times 2^2 =$ 15. $7^2 \times 4^2 =$

16. $2^4 \times 3^2 =$ 17. $6^2 \times 3^2 =$ 18. $4^3 \times 2^3 =$

19. $2^5 \times 2^2 =$ 20. $5^2 \times 2^3 =$ 21. $2^5 \times 3^3 =$

22. $2^5 \times 3^2 =$ 23. $3^3 \times 2^3 =$ 24. $4^2 \times 3^3 =$

25. $2^6 \times 2^2 =$ 26. $5^3 \times 2^3 =$ 27. $2^5 \times 2^5 =$

> Evaluate the following exponents.
>
> Example: $2^3 \times 3^2 = 8 \times 9 = 72$

1. $2^3 \times 4^2 =$

2. $2^4 \times 2^4 =$

3. $5^2 \times 2^6 =$

4. $3^3 \times 2^3 =$

5. $10^2 \times 2^3 =$

6. $2^5 \times 2^4 =$

7. $2^4 \times 4^2 =$

8. $11^2 \times 2^2 =$

9. $3^2 \times 2^5 =$

10. $5^2 \times 2^2 =$

11. $4^3 \times 2^4 =$

12. $4^2 \times 2^4 =$

13. $6^2 \times 2^3 =$

14. $2^5 \times 2^3 =$

15. $2^4 \times 5^3 =$

16. $3^3 \times 3^3 =$

17. $3^4 \times 2^3 =$

18. $2^3 \times 4^4 =$

19. $4^3 \times 2^2 =$

20. $2^6 \times 2^4 =$

21. $13^2 \times 2^2 =$

22. $5^3 \times 2^3 =$

23. $3^5 \times 2^2 =$

24. $3^5 \times 2^3 =$

25. $6^3 \times 2^2 =$

26. $6^3 \times 2^2 =$

27. $7^3 \times 2^2 =$

Lesson 1-1   Changing fractions to percents

Change the following fractions to percents.
Example: $\frac{1}{4} = 0.25 = 25\%$

1. $\frac{1}{4} = \underline{0.25} = \underline{25\%}$ 2. $\frac{1}{2} = \underline{0.5} = \underline{50\%}$

3. $\frac{1}{5} = \underline{0.2} = \underline{20\%}$ 4. $\frac{2}{5} = \underline{0.4} = \underline{40\%}$

5. $\frac{3}{4} = \underline{0.75} = \underline{75\%}$ 6. $\frac{4}{5} = \underline{0.8} = \underline{80\%}$

7. $\frac{3}{5} = \underline{0.6} = \underline{60\%}$ 8. $\frac{1}{10} = \underline{0.1} = \underline{10\%}$

9. $\frac{3}{10} = \underline{0.3} = \underline{30\%}$ 10. $\frac{2}{10} = \underline{0.2} = \underline{20\%}$

11. $\frac{1}{8} = \underline{0.125} = \underline{12.5\%}$ 12. $\frac{2}{8} = \underline{0.25} = \underline{25\%}$

13. $\frac{4}{8} = \underline{0.5} = \underline{50\%}$ 14. $\frac{3}{8} = \underline{0.375} = \underline{37.5\%}$

15. $\frac{5}{8} = \underline{0.625} = \underline{62.5\%}$ 16. $\frac{6}{8} = \underline{0.75} = \underline{75\%}$

Lesson 1-2   Changing fractions to percents

Change the following fractions to percents.
Example: $\frac{1}{4} = 0.25 = 25\%$

1. $\frac{2}{5} = \underline{0.4} = \underline{40\%}$ 2. $\frac{3}{5} = \underline{0.6} = \underline{60\%}$

3. $\frac{1}{4} = \underline{0.25} = \underline{25\%}$ 4. $\frac{3}{4} = \underline{0.75} = \underline{75\%}$

5. $\frac{3}{2} = \underline{1.5} = \underline{150\%}$ 6. $\frac{5}{2} = \underline{2.5} = \underline{250\%}$

7. $\frac{2}{10} = \underline{0.2} = \underline{20\%}$ 8. $\frac{1}{8} = \underline{0.125} = \underline{12.5\%}$

9. $\frac{5}{8} = \underline{0.625} = \underline{62.5\%}$ 10. $\frac{3}{10} = \underline{0.3} = \underline{30\%}$

11. $\frac{6}{10} = \underline{0.6} = \underline{60\%}$ 12. $\frac{4}{10} = \underline{0.4} = \underline{40\%}$

13. $\frac{6}{8} = \underline{0.75} = \underline{75\%}$ 14. $\frac{7}{8} = \underline{0.875} = \underline{87.5\%}$

15. $\frac{6}{4} = \underline{1.5} = \underline{150\%}$ 16. $\frac{12}{10} = \underline{1.2} = \underline{120\%}$

Lesson 1-3   Changing fractions to percents

Change the following fractions to percents.
Example: $\frac{1}{4} = 0.25 = 25\%$

1. $\frac{1}{8} = \underline{0.125} = \underline{12.5\%}$ 2. $\frac{3}{5} = \underline{0.6} = \underline{60\%}$

3. $\frac{7}{10} = \underline{0.7} = \underline{70\%}$ 4. $\frac{5}{4} = \underline{1.25} = \underline{125\%}$

5. $\frac{5}{2} = \underline{2.5} = \underline{250\%}$ 6. $\frac{9}{10} = \underline{0.9} = \underline{90\%}$

7. $\frac{3}{8} = \underline{0.375} = \underline{37.5\%}$ 8. $\frac{5}{8} = \underline{0.625} = \underline{62.5\%}$

9. $\frac{3}{4} = \underline{0.75} = \underline{75\%}$ 10. $\frac{3}{2} = \underline{1.5} = \underline{150\%}$

11. $\frac{4}{5} = \underline{0.8} = \underline{80\%}$ 12. $\frac{7}{8} = \underline{0.875} = \underline{87.5\%}$

13. $\frac{3}{20} = \underline{0.15} = \underline{15\%}$ 14. $\frac{1}{20} = \underline{0.05} = \underline{5\%}$

15. $\frac{13}{10} = \underline{1.3} = \underline{130\%}$ 16. $\frac{7}{4} = \underline{1.75} = \underline{175\%}$

Lesson 1-4   Changing fractions to percents

Change the following fractions to percents.
Example: $\frac{1}{4} = 0.25 = 25\%$

1. $\frac{9}{12} = \underline{0.75} = \underline{75\%}$ 2. $\frac{6}{10} = \underline{0.6} = \underline{60\%}$

3. $\frac{4}{10} = \underline{0.4} = \underline{40\%}$ 4. $\frac{6}{5} = \underline{1.2} = \underline{120\%}$

5. $\frac{7}{2} = \underline{3.5} = \underline{350\%}$ 6. $\frac{7}{4} = \underline{1.75} = \underline{175\%}$

7. $\frac{7}{5} = \underline{1.4} = \underline{140\%}$ 8. $\frac{3}{20} = \underline{0.15} = \underline{15\%}$

9. $\frac{12}{8} = \underline{1.5} = \underline{150\%}$ 10. $\frac{7}{8} = \underline{0.875} = \underline{87.5\%}$

11. $\frac{14}{10} = \underline{1.4} = \underline{140\%}$ 12. $\frac{7}{20} = \underline{0.35} = \underline{35\%}$

13. $\frac{5}{4} = \underline{1.25} = \underline{125\%}$ 14. $\frac{12}{20} = \underline{0.6} = \underline{60\%}$

15. $\frac{9}{20} = \underline{0.45} = \underline{45\%}$ 16. $\frac{11}{20} = \underline{0.55} = \underline{55\%}$

Change the following fractions to percents.
Example: $\frac{1}{4} = 0.25 = 25\%$

1. $\frac{1}{5} = \underline{0.2} = \underline{20\%}$ 2. $\frac{4}{8} = \underline{0.5} = \underline{50\%}$

3. $\frac{9}{10} = \underline{0.9} = \underline{90\%}$ 4. $\frac{7}{4} = \underline{1.75} = \underline{175\%}$

5. $\frac{9}{6} = \underline{1.5} = \underline{150\%}$ 6. $\frac{8}{5} = \underline{1.6} = \underline{160\%}$

7. $\frac{9}{15} = \underline{0.6} = \underline{60\%}$ 8. $\frac{7}{5} = \underline{1.4} = \underline{140\%}$

9. $\frac{4}{5} = \underline{1.25} = \underline{125\%}$ 10. $\frac{7}{10} = \underline{0.7} = \underline{70\%}$

11. $\frac{9}{20} = \underline{0.45} = \underline{45\%}$ 12. $\frac{7}{20} = \underline{0.35} = \underline{35\%}$

13. $\frac{13}{10} = \underline{1.3} = \underline{130\%}$ 14. $\frac{11}{10} = \underline{1.1} = \underline{110\%}$

15. $\frac{12}{8} = \underline{1.5} = \underline{150\%}$ 16. $\frac{13}{20} = \underline{0.65} = \underline{65\%}$

133

Change the following fractions to percents.
Example: $\frac{1}{4} = 0.25 = 25\%$

1. $\frac{14}{10} = \underline{1.4} = \underline{140\%}$ 2. $\frac{8}{10} = \underline{0.8} = \underline{80\%}$

3. $\frac{6}{2} = \underline{3} = \underline{300\%}$ 4. $\frac{5}{2} = \underline{2.5} = \underline{250\%}$

5. $\frac{7}{10} = \underline{0.7} = \underline{70\%}$ 6. $\frac{6}{10} = \underline{0.6} = \underline{60\%}$

7. $\frac{6}{20} = \underline{0.3} = \underline{30\%}$ 8. $\frac{3}{25} = \underline{0.12} = \underline{12\%}$

9. $\frac{9}{20} = \underline{0.45} = \underline{45\%}$ 10. $\frac{16}{10} = \underline{1.6} = \underline{160\%}$

11. $\frac{8}{5} = \underline{1.6} = \underline{160\%}$ 12. $\frac{7}{20} = \underline{0.35} = \underline{35\%}$

13. $\frac{15}{6} = \underline{2.5} = \underline{250\%}$ 14. $\frac{9}{5} = \underline{1.8} = \underline{180\%}$

15. $\frac{19}{20} = \underline{0.95} = \underline{95\%}$ 16. $\frac{17}{20} = \underline{0.85} = \underline{85\%}$

134

Change the following fractions to percents.
Example: $\frac{1}{4} = 0.25 = 25\%$

1. $\frac{7}{8} = \underline{0.875} = \underline{87.5\%}$ 2. $\frac{3}{8} = \underline{0.375} = \underline{37.5\%}$

3. $\frac{6}{10} = \underline{0.6} = \underline{60\%}$ 4. $\frac{9}{10} = \underline{0.9} = \underline{90\%}$

5. $\frac{16}{25} = \underline{0.64} = \underline{64\%}$ 6. $\frac{12}{10} = \underline{1.2} = \underline{120\%}$

7. $\frac{9}{8} = \underline{1.125} = \underline{112.5\%}$ 8. $\frac{11}{8} = \underline{1.375} = \underline{137.5\%}$

9. $\frac{7}{5} = \underline{1.4} = \underline{140\%}$ 10. $\frac{12}{5} = \underline{2.4} = \underline{240\%}$

11. $\frac{13}{20} = \underline{0.65} = \underline{65\%}$ 12. $\frac{7}{25} = \underline{0.28} = \underline{28\%}$

13. $\frac{20}{8} = \underline{2.5} = \underline{250\%}$ 14. $\frac{17}{20} = \underline{0.85} = \underline{85\%}$

15. $\frac{40}{20} = \underline{2} = \underline{200\%}$ 16. $\frac{22}{20} = \underline{1.1} = \underline{110\%}$

135

Change the following fractions to percents.
Example: $\frac{1}{4} = 0.25 = 25\%$

1. $\frac{8}{10} = \underline{0.8} = \underline{80\%}$ 2. $\frac{6}{4} = \underline{1.5} = \underline{150\%}$

3. $\frac{14}{10} = \underline{1.4} = \underline{140\%}$ 4. $\frac{8}{2} = \underline{4} = \underline{400\%}$

5. $\frac{9}{20} = \underline{0.45} = \underline{45\%}$ 6. $\frac{10}{8} = \underline{1.25} = \underline{125\%}$

7. $\frac{13}{8} = \underline{1.625} = \underline{162.5\%}$ 8. $\frac{17}{20} = \underline{0.85} = \underline{85\%}$

9. $\frac{24}{20} = \underline{1.2} = \underline{120\%}$ 10. $\frac{30}{20} = \underline{1.5} = \underline{150\%}$

11. $\frac{13}{5} = \underline{2.6} = \underline{260\%}$ 12. $\frac{16}{5} = \underline{3.2} = \underline{320\%}$

13. $\frac{3}{20} = \underline{0.15} = \underline{15\%}$ 14. $\frac{15}{8} = \underline{1.875} = \underline{187.5\%}$

15. $\frac{9}{25} = \underline{0.36} = \underline{36\%}$ 16. $\frac{11}{25} = \underline{0.44} = \underline{44\%}$

136

Change the following fractions to percents.

Example: $\frac{1}{4} = 0.25 = 25\%$

1. $\frac{9}{4} = \underline{2.25} = \underline{225\%}$ 2. $\frac{3}{8} = \underline{0.375} = \underline{37.5\%}$

3. $\frac{6}{10} = \underline{0.6} = \underline{60\%}$ 4. $\frac{4}{5} = \underline{0.8} = \underline{80\%}$

5. $\frac{25}{10} = \underline{2.5} = \underline{250\%}$ 6. $\frac{5}{4} = \underline{1.25} = \underline{125\%}$

7. $\frac{20}{8} = \underline{2.5} = \underline{250\%}$ 8. $\frac{11}{8} = \underline{1.375} = \underline{137.5\%}$

9. $\frac{7}{20} = \underline{0.35} = \underline{35\%}$ 10. $\frac{15}{20} = \underline{0.75} = \underline{75\%}$

11. $\frac{4}{25} = \underline{0.16} = \underline{16\%}$ 12. $\frac{35}{10} = \underline{3.5} = \underline{350\%}$

13. $\frac{17}{10} = \underline{1.7} = \underline{170\%}$ 14. $\frac{36}{20} = \underline{1.8} = \underline{180\%}$

15. $\frac{40}{20} = \underline{2} = \underline{200\%}$ 16. $\frac{17}{5} = \underline{3.4} = \underline{340\%}$

Change the following fractions to percents.

Example: $\frac{1}{4} = 0.25 = 25\%$

1. $\frac{11}{10} = \underline{1.1} = \underline{110\%}$ 2. $\frac{6}{20} = \underline{0.3} = \underline{30\%}$

3. $\frac{8}{2} = \underline{4} = \underline{400\%}$ 4. $\frac{9}{2} = \underline{4.5} = \underline{450\%}$

5. $\frac{9}{8} = \underline{1.125} = \underline{112.5\%}$ 6. $\frac{9}{6} = \underline{1.5} = \underline{150\%}$

7. $\frac{12}{8} = \underline{1.5} = \underline{150\%}$ 8. $\frac{25}{10} = \underline{2.5} = \underline{250\%}$

9. $\frac{7}{25} = \underline{0.28} = \underline{28\%}$ 10. $\frac{14}{5} = \underline{2.8} = \underline{280\%}$

11. $\frac{38}{20} = \underline{1.9} = \underline{190\%}$ 12. $\frac{39}{20} = \underline{1.95} = \underline{195\%}$

13. $\frac{18}{5} = \underline{1.875} = \underline{187.5\%}$ 14. $\frac{17}{20} = \underline{0.85} = \underline{85\%}$

15. $\frac{11}{20} = \underline{0.55} = \underline{55\%}$ 16. $\frac{18}{6} = \underline{3} = \underline{300\%}$

Change the following percents to fractions.

Example: $75\% = \frac{75}{100} = \frac{3}{4}$

1. $25\% = \frac{25}{100} = \frac{1}{4}$ 2. $20\% = \frac{20}{100} = \frac{1}{5}$

3. $50\% = \frac{50}{100} = \frac{1}{2}$ 4. $30\% = \frac{30}{100} = \frac{3}{10}$

5. $35\% = \frac{35}{100} = \frac{7}{20}$ 6. $45\% = \frac{45}{100} = \frac{9}{25}$

7. $60\% = \frac{60}{100} = \frac{3}{5}$ 8. $55\% = \frac{55}{100} = \frac{11}{20}$

9. $40\% = \frac{40}{100} = \frac{2}{5}$ 10. $70\% = \frac{70}{100} = \frac{7}{10}$

11. $4\% = \frac{4}{100} = \frac{1}{25}$ 12. $12\% = \frac{12}{100} = \frac{3}{25}$

13. $32\% = \frac{32}{100} = \frac{8}{25}$ 14. $80\% = \frac{80}{100} = \frac{4}{5}$

15. $90\% = \frac{90}{100} = \frac{9}{10}$ 16. $96\% = \frac{96}{100} = \frac{24}{25}$

Change the following percents to fractions.

Example: $75\% = \frac{75}{100} = \frac{3}{4}$

1. $50\% = \frac{50}{100} = \frac{1}{2}$ 2. $85\% = \frac{85}{100} = \frac{17}{20}$

3. $35\% = \frac{35}{100} = \frac{7}{20}$ 4. $15\% = \frac{15}{100} = \frac{3}{20}$

5. $80\% = \frac{80}{100} = \frac{4}{5}$ 6. $75\% = \frac{75}{100} = \frac{3}{4}$

7. $45\% = \frac{45}{100} = \frac{9}{20}$ 8. $90\% = \frac{90}{100} = \frac{9}{10}$

9. $44\% = \frac{44}{100} = \frac{11}{25}$ 10. $55\% = \frac{55}{100} = \frac{11}{20}$

11. $95\% = \frac{95}{100} = \frac{19}{20}$ 12. $56\% = \frac{56}{100} = \frac{14}{25}$

13. $72\% = \frac{72}{100} = \frac{18}{25}$ 14. $48\% = \frac{48}{100} = \frac{12}{25}$

15. $65\% = \frac{65}{100} = \frac{13}{20}$ 16. $96\% = \frac{96}{100} = \frac{24}{25}$

Change the following percents to fractions.
Example: $75\% = \frac{75}{100} = \frac{3}{4}$

1. $25\% = \frac{25}{100} = \frac{1}{4}$ 2. $50\% = \frac{50}{100} = \frac{1}{2}$

3. $16\% = \frac{16}{100} = \frac{4}{25}$ 4. $70\% = \frac{70}{100} = \frac{7}{10}$

5. $45\% = \frac{45}{100} = \frac{9}{20}$ 6. $65\% = \frac{65}{100} = \frac{13}{20}$

7. $42\% = \frac{42}{100} = \frac{21}{50}$ 8. $35\% = \frac{35}{100} = \frac{7}{20}$

9. $90\% = \frac{90}{100} = \frac{9}{10}$ 10. $88\% = \frac{88}{100} = \frac{22}{25}$

11. $20\% = \frac{20}{100} = \frac{1}{5}$ 12. $64\% = \frac{64}{100} = \frac{16}{25}$

13. $92\% = \frac{92}{100} = \frac{23}{25}$ 14. $8\% = \frac{8}{100} = \frac{2}{25}$

15. $95\% = \frac{95}{100} = \frac{19}{20}$ 16. $15\% = \frac{15}{100} = \frac{3}{20}$

Change the following percents to fractions.
Example: $75\% = \frac{75}{100} = \frac{3}{4}$

1. $82\% = \frac{82}{100} = \frac{41}{50}$ 2. $35\% = \frac{35}{100} = \frac{7}{20}$

3. $125\% = \frac{125}{100} = \frac{5}{4}$ 4. $150\% = \frac{150}{100} = \frac{3}{2}$

5. $15\% = \frac{15}{100} = \frac{3}{20}$ 6. $45\% = \frac{45}{100} = \frac{9}{20}$

7. $250\% = \frac{250}{100} = \frac{5}{2}$ 8. $64\% = \frac{64}{100} = \frac{16}{25}$

9. $72\% = \frac{72}{100} = \frac{18}{25}$ 10. $175\% = \frac{175}{100} = \frac{7}{4}$

11. $55\% = \frac{55}{100} = \frac{11}{20}$ 12. $65\% = \frac{65}{100} = \frac{13}{20}$

13. $170\% = \frac{170}{100} = \frac{17}{10}$ 14. $275\% = \frac{275}{100} = \frac{11}{4}$

15. $24\% = \frac{24}{100} = \frac{6}{25}$ 16. $34\% = \frac{34}{100} = \frac{17}{50}$

Change the following percents to fractions.
Example: $75\% = \frac{75}{100} = \frac{3}{4}$

1. $175\% = \frac{175}{100} = \frac{7}{4}$ 2. $120\% = \frac{120}{100} = \frac{6}{5}$

3. $48\% = \frac{48}{100} = \frac{12}{25}$ 4. $65\% = \frac{65}{100} = \frac{13}{20}$

5. $110\% = \frac{110}{100} = \frac{11}{10}$ 6. $5\% = \frac{5}{100} = \frac{1}{20}$

7. $14\% = \frac{14}{100} = \frac{7}{50}$ 8. $8\% = \frac{8}{100} = \frac{4}{25}$

9. $92\% = \frac{92}{100} = \frac{23}{25}$ 10. $180\% = \frac{180}{100} = \frac{9}{5}$

11. $190\% = \frac{190}{100} = \frac{19}{10}$ 12. $325\% = \frac{325}{100} = \frac{13}{4}$

13. $64\% = \frac{64}{100} = \frac{16}{25}$ 14. $210\% = \frac{210}{100} = \frac{21}{10}$

15. $24\% = \frac{24}{100} = \frac{6}{25}$ 16. $76\% = \frac{76}{100} = \frac{19}{25}$

Change the following percents to fractions.
Example: $75\% = \frac{75}{100} = \frac{3}{4}$

1. $98\% = \frac{98}{100} = \frac{49}{50}$ 2. $130\% = \frac{130}{100} = \frac{13}{10}$

3. $88\% = \frac{88}{100} = \frac{22}{25}$ 4. $37.5\% = \frac{37.5}{100} = \frac{375}{1000} = \frac{3}{8}$

5. $225\% = \frac{225}{100} = \frac{9}{4}$ 6. $150\% = \frac{150}{100} = \frac{3}{2}$

7. $12.5\% = \frac{12.5}{100} = \frac{125}{1000} = \frac{1}{8}$ 8. $44\% = \frac{44}{100} = \frac{11}{25}$

9. $175\% = \frac{175}{100} = \frac{7}{4}$ 10. $62.5\% = \frac{62.5}{100} = \frac{625}{1000} = \frac{5}{8}$

11. $155\% = \frac{155}{100} = \frac{31}{20}$ 12. $125\% = \frac{125}{100} = \frac{5}{4}$

13. $87.5\% = \frac{87.5}{100} = \frac{875}{1000} = \frac{7}{8}$ 14. $165\% = \frac{165}{100} = \frac{33}{20}$

15. $105\% = \frac{105}{100} = \frac{21}{20}$ 16. $220\% = \frac{220}{100} = \frac{11}{5}$

Change the following percents to fractions.

Example: $75\% = \frac{75}{100} = \frac{3}{4}$

1. $62.5\% = \dfrac{62.5}{100} = \dfrac{625}{1000} = \dfrac{5}{8}$ 2. $65\% = \dfrac{65}{100} = \dfrac{13}{20}$

3. $170\% = \dfrac{170}{100} = \dfrac{17}{10}$ 4. $140\% = \dfrac{140}{100} = \dfrac{7}{5}$

5. $200\% = \dfrac{200}{100} = 2$ 6. $87.5\% = \dfrac{87.5}{100} = \dfrac{875}{1000} = \dfrac{7}{8}$

7. $145\% = \dfrac{145}{100} = \dfrac{29}{20}$ 8. $5\% = \dfrac{5}{100} = \dfrac{1}{20}$

9. $95\% = \dfrac{95}{100} = \dfrac{19}{20}$ 10. $225\% = \dfrac{225}{100} = \dfrac{9}{4}$

11. $144\% = \dfrac{144}{100} = \dfrac{36}{25}$ 12. $155\% = \dfrac{155}{100} = \dfrac{31}{20}$

13. $12.5\% = \dfrac{12.5}{100} = \dfrac{125}{1000} = \dfrac{1}{8}$ 14. $124\% = \dfrac{124}{100} = \dfrac{31}{25}$

15. $54\% = \dfrac{54}{100} = \dfrac{27}{50}$ 16. $195\% = \dfrac{195}{100} = \dfrac{39}{20}$

Change the following percents to fractions.

Example: $75\% = \frac{75}{100} = \frac{3}{4}$

1. $28\% = \dfrac{28}{100} = \dfrac{7}{25}$ 2. $52\% = \dfrac{52}{100} = \dfrac{13}{25}$

3. $35\% = \dfrac{35}{100} = \dfrac{7}{20}$ 4. $36\% = \dfrac{36}{100} = \dfrac{9}{25}$

5. $190\% = \dfrac{190}{100} = \dfrac{19}{10}$ 6. $85\% = \dfrac{85}{100} = \dfrac{17}{20}$

7. $250\% = \dfrac{250}{100} = \dfrac{5}{2}$ 8. $128\% = \dfrac{128}{100} = \dfrac{32}{25}$

9. $175\% = \dfrac{175}{100} = \dfrac{7}{4}$ 10. $300\% = \dfrac{300}{100} = 3$

11. $115\% = \dfrac{115}{100} = \dfrac{23}{20}$ 12. $225\% = \dfrac{225}{100} = \dfrac{9}{4}$

13. $108\% = \dfrac{108}{100} = \dfrac{27}{25}$ 14. $188\% = \dfrac{188}{100} = \dfrac{47}{25}$

15. $280\% = \dfrac{280}{100} = \dfrac{14}{5}$ 16. $152\% = \dfrac{152}{100} = \dfrac{38}{25}$

Change the following percents to fractions.

Example: $75\% = \frac{75}{100} = \frac{3}{4}$

1. $56\% = \dfrac{56}{100} = \dfrac{14}{25}$ 2. $250\% = \dfrac{250}{100} = \dfrac{5}{2}$

3. $130\% = \dfrac{130}{100} = \dfrac{13}{10}$ 4. $64\% = \dfrac{64}{100} = \dfrac{16}{25}$

5. $325\% = \dfrac{325}{100} = \dfrac{13}{4}$ 6. $45\% = \dfrac{45}{100} = \dfrac{9}{20}$

7. $112\% = \dfrac{112}{100} = \dfrac{28}{25}$ 8. $96\% = \dfrac{96}{100} = \dfrac{24}{25}$

9. $35\% = \dfrac{35}{100} = \dfrac{7}{20}$ 10. $102\% = \dfrac{102}{100} = \dfrac{51}{50}$

11. $165\% = \dfrac{165}{100} = \dfrac{33}{20}$ 12. $375\% = \dfrac{375}{100} = \dfrac{15}{4}$

13. $148\% = \dfrac{148}{100} = \dfrac{37}{25}$ 14. $16\% = \dfrac{16}{100} = \dfrac{4}{25}$

15. $62.5\% = \dfrac{62.5}{100} = \dfrac{625}{1000} = \dfrac{5}{8}$ 16. $152\% = \dfrac{152}{100} = \dfrac{38}{25}$

Change the following percents to fractions.

Example: $75\% = \frac{75}{100} = \frac{3}{4}$

1. $275\% = \dfrac{275}{100} = \dfrac{11}{4}$ 2. $375\% = \dfrac{375}{100} = \dfrac{15}{4}$

3. $146\% = \dfrac{146}{100} = \dfrac{73}{50}$ 4. $220\% = \dfrac{220}{100} = \dfrac{11}{5}$

5. $115\% = \dfrac{115}{100} = \dfrac{23}{20}$ 6. $96\% = \dfrac{96}{100} = \dfrac{24}{25}$

7. $196\% = \dfrac{196}{100} = \dfrac{49}{25}$ 8. $164\% = \dfrac{164}{100} = \dfrac{41}{25}$

9. $400\% = \dfrac{400}{100} = 4$ 10. $138\% = \dfrac{138}{100} = \dfrac{69}{50}$

11. $150\% = \dfrac{150}{100} = \dfrac{3}{2}$ 12. $105\% = \dfrac{105}{100} = \dfrac{21}{20}$

13. $215\% = \dfrac{215}{100} = \dfrac{43}{20}$ 14. $172\% = \dfrac{172}{100} = \dfrac{43}{25}$

15. $185\% = \dfrac{185}{100} = \dfrac{37}{20}$ 16. $350\% = \dfrac{350}{100} = \dfrac{7}{2}$

Change the following mixed numbers to improper fractions.

Example: $2\frac{1}{2} = 2 + \frac{1}{2} = \frac{4}{2} + \frac{1}{2} = \frac{5}{2}$

1. $1\frac{2}{3} = \frac{5}{3}$ 2. $1\frac{1}{4} = \frac{5}{4}$ 3. $2\frac{1}{5} = \frac{11}{5}$

4. $2\frac{2}{5} = \frac{12}{5}$ 5. $1\frac{1}{3} = \frac{4}{3}$ 6. $2\frac{2}{3} = \frac{8}{3}$

7. $1\frac{2}{7} = \frac{9}{7}$ 8. $1\frac{3}{7} = \frac{10}{7}$ 9. $2\frac{3}{7} = \frac{17}{7}$

10. $1\frac{3}{4} = \frac{7}{4}$ 11. $2\frac{3}{5} = \frac{13}{5}$ 12. $1\frac{3}{5} = \frac{8}{5}$

13. $2\frac{4}{5} = \frac{14}{5}$ 14. $2\frac{2}{7} = \frac{16}{7}$ 15. $1\frac{6}{7} = \frac{13}{7}$

16. $1\frac{1}{6} = \frac{7}{6}$ 17. $1\frac{4}{7} = \frac{11}{7}$ 18. $2\frac{1}{7} = \frac{15}{7}$

19. $2\frac{5}{7} = \frac{19}{7}$ 20. $1\frac{5}{6} = \frac{11}{6}$ 21. $3\frac{5}{8} = \frac{29}{8}$

22. $1\frac{7}{8} = \frac{15}{8}$ 23. $2\frac{5}{8} = \frac{21}{8}$ 24. $3\frac{5}{6} = \frac{23}{6}$

Change the following mixed numbers to improper fractions.

Example: $2\frac{1}{2} = 2 + \frac{1}{2} = \frac{4}{2} + \frac{1}{2} = \frac{5}{2}$

1. $1\frac{2}{3} = \frac{5}{3}$ 2. $1\frac{1}{6} = \frac{7}{6}$ 3. $2\frac{4}{5} = \frac{14}{5}$

4. $2\frac{1}{4} = \frac{9}{4}$ 5. $1\frac{1}{3} = \frac{4}{3}$ 6. $3\frac{1}{2} = \frac{7}{2}$

7. $1\frac{2}{5} = \frac{7}{5}$ 8. $1\frac{3}{4} = \frac{7}{4}$ 9. $2\frac{4}{9} = \frac{22}{9}$

10. $1\frac{3}{7} = \frac{10}{7}$ 11. $2\frac{3}{5} = \frac{13}{5}$ 12. $1\frac{6}{7} = \frac{13}{7}$

13. $2\frac{5}{6} = \frac{17}{6}$ 14. $2\frac{4}{7} = \frac{18}{7}$ 15. $1\frac{5}{9} = \frac{14}{9}$

16. $3\frac{1}{8} = \frac{25}{8}$ 17. $3\frac{3}{8} = \frac{27}{8}$ 18. $2\frac{7}{10} = \frac{27}{10}$

19. $2\frac{1}{9} = \frac{19}{9}$ 20. $1\frac{2}{9} = \frac{11}{9}$ 21. $2\frac{5}{8} = \frac{21}{8}$

22. $3\frac{3}{10} = \frac{33}{10}$ 23. $3\frac{7}{8} = \frac{31}{8}$ 24. $3\frac{5}{7} = \frac{26}{7}$

Change the following mixed numbers to improper fractions.

Example: $2\frac{1}{2} = 2 + \frac{1}{2} = \frac{4}{2} + \frac{1}{2} = \frac{5}{2}$

1. $3\frac{1}{6} = \frac{19}{6}$ 2. $2\frac{5}{6} = \frac{17}{6}$ 3. $1\frac{6}{7} = \frac{13}{7}$

4. $3\frac{3}{4} = \frac{15}{4}$ 5. $3\frac{1}{4} = \frac{13}{4}$ 6. $3\frac{3}{10} = \frac{33}{10}$

7. $1\frac{1}{10} = \frac{11}{10}$ 8. $2\frac{5}{6} = \frac{17}{6}$ 9. $2\frac{4}{9} = \frac{22}{9}$

10. $2\frac{3}{4} = \frac{11}{4}$ 11. $2\frac{5}{8} = \frac{21}{8}$ 12. $1\frac{8}{9} = \frac{17}{9}$

13. $2\frac{3}{8} = \frac{19}{8}$ 14. $3\frac{7}{10} = \frac{37}{10}$ 15. $2\frac{6}{7} = \frac{20}{7}$

16. $2\frac{1}{8} = \frac{17}{8}$ 17. $1\frac{5}{7} = \frac{12}{7}$ 18. $3\frac{1}{8} = \frac{25}{8}$

19. $3\frac{1}{9} = \frac{28}{9}$ 20. $2\frac{7}{8} = \frac{23}{8}$ 21. $3\frac{2}{7} = \frac{23}{7}$

22. $1\frac{9}{10} = \frac{19}{10}$ 23. $2\frac{7}{10} = \frac{27}{10}$ 24. $3\frac{1}{10} = \frac{31}{10}$

Change the following mixed numbers to improper fractions.

Example: $2\frac{1}{2} = 2 + \frac{1}{2} = \frac{4}{2} + \frac{1}{2} = \frac{5}{2}$

1. $2\frac{1}{5} = \frac{11}{5}$ 2. $1\frac{5}{6} = \frac{11}{6}$ 3. $3\frac{4}{5} = \frac{19}{5}$

4. $4\frac{3}{4} = \frac{19}{4}$ 5. $3\frac{1}{4} = \frac{13}{4}$ 6. $2\frac{3}{7} = \frac{17}{7}$

7. $3\frac{1}{6} = \frac{19}{6}$ 8. $2\frac{2}{5} = \frac{12}{5}$ 9. $3\frac{5}{8} = \frac{29}{8}$

10. $3\frac{2}{7} = \frac{23}{7}$ 11. $2\frac{1}{7} = \frac{15}{7}$ 12. $4\frac{4}{7} = \frac{32}{7}$

13. $1\frac{3}{8} = \frac{11}{8}$ 14. $4\frac{3}{8} = \frac{35}{8}$ 15. $2\frac{5}{9} = \frac{23}{9}$

16. $4\frac{2}{9} = \frac{38}{9}$ 17. $3\frac{4}{9} = \frac{31}{9}$ 18. $3\frac{6}{7} = \frac{27}{7}$

19. $3\frac{3}{11} = \frac{36}{11}$ 20. $4\frac{3}{10} = \frac{43}{10}$ 21. $2\frac{7}{11} = \frac{29}{11}$

22. $3\frac{7}{10} = \frac{37}{10}$ 23. $4\frac{5}{7} = \frac{33}{7}$ 24. $4\frac{7}{8} = \frac{39}{8}$

Change the following mixed numbers to improper fractions.

Example: $2\frac{1}{2} = 2 + \frac{1}{2} = \frac{4}{2} + \frac{1}{2} = \frac{5}{2}$

1. $2\frac{1}{4} = \frac{9}{4}$ 2. $3\frac{1}{6} = \frac{19}{6}$ 3. $2\frac{6}{7} = \frac{20}{7}$

4. $4\frac{5}{6} = \frac{29}{6}$ 5. $2\frac{5}{7} = \frac{19}{7}$ 6. $3\frac{5}{9} = \frac{32}{9}$

7. $3\frac{3}{5} = \frac{18}{5}$ 8. $4\frac{2}{5} = \frac{22}{5}$ 9. $5\frac{4}{5} = \frac{29}{5}$

10. $5\frac{2}{7} = \frac{37}{7}$ 11. $4\frac{3}{4} = \frac{19}{4}$ 12. $4\frac{7}{9} = \frac{43}{9}$

13. $2\frac{4}{9} = \frac{22}{9}$ 14. $5\frac{2}{9} = \frac{47}{9}$ 15. $5\frac{9}{10} = \frac{59}{10}$

16. $3\frac{8}{9} = \frac{35}{9}$ 17. $2\frac{3}{10} = \frac{23}{10}$ 18. $3\frac{3}{11} = \frac{36}{11}$

19. $5\frac{7}{10} = \frac{57}{10}$ 20. $3\frac{2}{11} = \frac{35}{11}$ 21. $4\frac{5}{12} = \frac{53}{12}$

22. $4\frac{4}{11} = \frac{48}{11}$ 23. $5\frac{1}{12} = \frac{61}{12}$ 24. $5\frac{1}{13} = \frac{66}{13}$

Change the following mixed numbers to improper fractions.

Example: $2\frac{1}{2} = 2 + \frac{1}{2} = \frac{4}{2} + \frac{1}{2} = \frac{5}{2}$

1. $2\frac{2}{13} = \frac{28}{13}$ 2. $3\frac{5}{8} = \frac{29}{8}$ 3. $3\frac{4}{9} = \frac{31}{9}$

4. $4\frac{7}{8} = \frac{39}{8}$ 5. $2\frac{2}{5} = \frac{12}{5}$ 6. $4\frac{5}{6} = \frac{29}{6}$

7. $5\frac{1}{5} = \frac{26}{5}$ 8. $3\frac{3}{13} = \frac{42}{13}$ 9. $5\frac{4}{5} = \frac{29}{5}$

10. $6\frac{1}{4} = \frac{25}{4}$ 11. $5\frac{2}{9} = \frac{47}{9}$ 12. $4\frac{3}{8} = \frac{35}{8}$

13. $4\frac{3}{10} = \frac{43}{10}$ 14. $5\frac{7}{10} = \frac{57}{10}$ 15. $6\frac{3}{4} = \frac{27}{4}$

16. $5\frac{5}{9} = \frac{50}{9}$ 17. $4\frac{7}{9} = \frac{43}{9}$ 18. $5\frac{7}{13} = \frac{72}{13}$

19. $3\frac{10}{11} = \frac{43}{11}$ 20. $5\frac{9}{11} = \frac{64}{11}$ 21. $6\frac{9}{10} = \frac{69}{10}$

22. $4\frac{1}{14} = \frac{57}{14}$ 23. $3\frac{3}{14} = \frac{45}{14}$ 24. $4\frac{8}{11} = \frac{52}{11}$

Change the following mixed numbers to improper fractions.

Example: $2\frac{1}{2} = 2 + \frac{1}{2} = \frac{4}{2} + \frac{1}{2} = \frac{5}{2}$

1. $3\frac{3}{7} = \frac{24}{7}$ 2. $4\frac{2}{3} = \frac{14}{3}$ 3. $4\frac{3}{4} = \frac{19}{4}$

4. $5\frac{5}{6} = \frac{35}{6}$ 5. $6\frac{7}{10} = \frac{67}{10}$ 6. $6\frac{2}{5} = \frac{32}{5}$

7. $4\frac{3}{10} = \frac{43}{10}$ 8. $5\frac{6}{7} = \frac{41}{7}$ 9. $3\frac{4}{7} = \frac{25}{7}$

10. $5\frac{3}{5} = \frac{28}{5}$ 11. $5\frac{4}{5} = \frac{29}{5}$ 12. $5\frac{4}{11} = \frac{59}{11}$

13. $6\frac{2}{9} = \frac{56}{9}$ 14. $4\frac{5}{9} = \frac{41}{9}$ 15. $4\frac{11}{12} = \frac{59}{12}$

16. $4\frac{5}{12} = \frac{53}{12}$ 17. $6\frac{2}{11} = \frac{68}{11}$ 18. $6\frac{5}{7} = \frac{47}{7}$

19. $4\frac{7}{11} = \frac{51}{11}$ 20. $3\frac{9}{13} = \frac{48}{13}$ 21. $6\frac{4}{9} = \frac{58}{9}$

22. $6\frac{4}{13} = \frac{82}{13}$ 23. $4\frac{7}{12} = \frac{55}{12}$ 24. $5\frac{8}{13} = \frac{73}{13}$

Change the following mixed numbers to improper fractions.

Example: $2\frac{1}{2} = 2 + \frac{1}{2} = \frac{4}{2} + \frac{1}{2} = \frac{5}{2}$

1. $6\frac{3}{8} = \frac{51}{8}$ 2. $6\frac{3}{5} = \frac{33}{5}$ 3. $6\frac{1}{6} = \frac{37}{6}$

4. $5\frac{1}{6} = \frac{31}{6}$ 5. $4\frac{2}{11} = \frac{46}{11}$ 6. $5\frac{3}{11} = \frac{58}{11}$

7. $7\frac{2}{5} = \frac{37}{5}$ 8. $5\frac{5}{8} = \frac{45}{8}$ 9. $6\frac{4}{5} = \frac{34}{5}$

10. $6\frac{2}{7} = \frac{44}{7}$ 11. $4\frac{4}{7} = \frac{32}{7}$ 12. $4\frac{5}{14} = \frac{61}{14}$

13. $5\frac{9}{14} = \frac{79}{14}$ 14. $7\frac{3}{4} = \frac{31}{4}$ 15. $7\frac{7}{8} = \frac{63}{8}$

16. $8\frac{2}{3} = \frac{26}{3}$ 17. $5\frac{5}{6} = \frac{35}{6}$ 18. $6\frac{7}{11} = \frac{73}{11}$

19. $5\frac{11}{12} = \frac{71}{12}$ 20. $4\frac{2}{13} = \frac{54}{13}$ 21. $5\frac{3}{7} = \frac{38}{7}$

22. $7\frac{8}{15} = \frac{113}{15}$ 23. $7\frac{7}{15} = \frac{112}{15}$ 24. $7\frac{3}{14} = \frac{101}{14}$

Change the following mixed numbers to improper fractions.

Example: $2\frac{1}{2} = 2 + \frac{1}{2} = \frac{4}{2} + \frac{1}{2} = \frac{5}{2}$

1. $5\frac{1}{7} = \frac{36}{7}$ 2. $5\frac{8}{11} = \frac{63}{11}$ 3. $8\frac{1}{6} = \frac{49}{6}$

4. $6\frac{2}{5} = \frac{32}{5}$ 5. $6\frac{2}{3} = \frac{20}{3}$ 6. $6\frac{7}{12} = \frac{79}{12}$

7. $6\frac{4}{11} = \frac{70}{11}$ 8. $5\frac{5}{7} = \frac{40}{7}$ 9. $5\frac{4}{13} = \frac{69}{13}$

10. $7\frac{5}{6} = \frac{47}{6}$ 11. $7\frac{8}{9} = \frac{71}{9}$ 12. $6\frac{11}{15} = \frac{101}{15}$

13. $5\frac{7}{16} = \frac{87}{16}$ 14. $8\frac{3}{5} = \frac{43}{5}$ 15. $8\frac{3}{4} = \frac{35}{4}$

16. $8\frac{3}{7} = \frac{59}{7}$ 17. $6\frac{6}{15} = \frac{96}{15}$ 18. $6\frac{7}{13} = \frac{85}{13}$

19. $9\frac{7}{9} = \frac{88}{9}$ 20. $7\frac{5}{14} = \frac{103}{14}$ 21. $7\frac{6}{7} = \frac{55}{7}$

22. $8\frac{8}{17} = \frac{144}{17}$ 23. $9\frac{9}{16} = \frac{153}{16}$ 24. $9\frac{4}{5} = \frac{49}{5}$

Change the following mixed numbers to improper fractions.

Example: $2\frac{1}{2} = 2 + \frac{1}{2} = \frac{4}{2} + \frac{1}{2} = \frac{5}{2}$

1. $8\frac{3}{4} = \frac{35}{4}$ 2. $8\frac{2}{7} = \frac{58}{7}$ 3. $8\frac{7}{10} = \frac{87}{10}$

4. $7\frac{5}{6} = \frac{47}{6}$ 5. $10\frac{2}{3} = \frac{32}{3}$ 6. $9\frac{3}{5} = \frac{48}{5}$

7. $9\frac{4}{5} = \frac{49}{5}$ 8. $7\frac{9}{20} = \frac{149}{20}$ 9. $7\frac{5}{12} = \frac{89}{12}$

10. $10\frac{5}{8} = \frac{85}{8}$ 11. $9\frac{8}{11} = \frac{107}{11}$ 12. $9\frac{11}{20} = \frac{191}{20}$

13. $8\frac{13}{18} = \frac{157}{18}$ 14. $9\frac{6}{7} = \frac{69}{7}$ 15. $10\frac{2}{5} = \frac{52}{5}$

16. $8\frac{9}{16} = \frac{137}{16}$ 17. $7\frac{4}{19} = \frac{137}{19}$ 18. $9\frac{8}{17} = \frac{161}{17}$

19. $9\frac{9}{19} = \frac{180}{19}$ 20. $8\frac{8}{15} = \frac{128}{15}$ 21. $10\frac{4}{7} = \frac{74}{7}$

22. $10\frac{17}{20} = \frac{217}{20}$ 23. $10\frac{7}{8} = \frac{87}{8}$ 24. $8\frac{9}{13} = \frac{113}{13}$

Change the following mixed numbers to improper fractions.

Example: $\frac{13}{5} = \frac{10}{5} + \frac{3}{5} = 2 + \frac{3}{5} = 2\frac{3}{5}$

1. $\frac{11}{3} = 3\frac{2}{3}$ 2. $\frac{9}{5} = 1\frac{4}{5}$ 3. $\frac{5}{3} = 1\frac{2}{3}$

4. $\frac{7}{4} = 1\frac{3}{4}$ 5. $\frac{10}{3} = 3\frac{1}{3}$ 6. $\frac{11}{5} = 2\frac{1}{5}$

7. $\frac{14}{5} = 2\frac{4}{5}$ 8. $\frac{13}{6} = 2\frac{1}{6}$ 9. $\frac{13}{3} = 4\frac{1}{3}$

10. $\frac{11}{8} = 1\frac{3}{8}$ 11. $\frac{13}{4} = 3\frac{1}{4}$ 12. $\frac{8}{6} = 1\frac{1}{3}$

13. $\frac{11}{9} = 1\frac{2}{9}$ 14. $\frac{9}{7} = 1\frac{2}{7}$ 15. $\frac{15}{8} = 1\frac{7}{8}$

16. $\frac{9}{6} = 1\frac{1}{2}$ 17. $\frac{10}{8} = 1\frac{1}{4}$ 18. $\frac{11}{7} = 1\frac{4}{7}$

19. $\frac{13}{10} = 1\frac{3}{10}$ 20. $\frac{13}{9} = 1\frac{4}{9}$ 21. $\frac{19}{9} = 2\frac{1}{9}$

22. $\frac{14}{11} = 1\frac{3}{11}$ 23. $\frac{16}{12} = 1\frac{1}{3}$ 24. $\frac{15}{12} = 1\frac{1}{4}$

Change the following improper fractions to mixed numbers.

Example: $\frac{13}{5} = \frac{10}{5} + \frac{3}{5} = 2 + \frac{3}{5} = 2\frac{3}{5}$

1. $\frac{17}{5} = 3\frac{2}{5}$ 2. $\frac{10}{7} = 1\frac{3}{10}$ 3. $\frac{7}{6} = 1\frac{1}{6}$

4. $\frac{6}{4} = 1\frac{1}{2}$ 5. $\frac{15}{2} = 7\frac{1}{2}$ 6. $\frac{19}{5} = 3\frac{4}{5}$

7. $\frac{13}{7} = 1\frac{6}{7}$ 8. $\frac{11}{4} = 2\frac{3}{4}$ 9. $\frac{13}{3} = 4\frac{1}{3}$

10. $\frac{19}{8} = 2\frac{3}{8}$ 11. $\frac{29}{5} = 5\frac{4}{5}$ 12. $\frac{29}{6} = 4\frac{5}{6}$

13. $\frac{20}{9} = 2\frac{2}{9}$ 14. $\frac{21}{8} = 2\frac{5}{8}$ 15. $\frac{15}{7} = 2\frac{1}{7}$

16. $\frac{39}{10} = 3\frac{9}{10}$ 17. $\frac{33}{10} = 3\frac{3}{10}$ 18. $\frac{44}{9} = 4\frac{8}{9}$

19. $\frac{27}{12} = 2\frac{1}{4}$ 20. $\frac{31}{9} = 3\frac{4}{9}$ 21. $\frac{26}{11} = 2\frac{4}{11}$

22. $\frac{20}{11} = 1\frac{9}{11}$ 23. $\frac{38}{11} = 3\frac{5}{11}$ 24. $\frac{47}{10} = 4\frac{7}{10}$

Change the following improper fractions to mixed numbers.

Example: $\dfrac{13}{5} = \dfrac{10}{5} + \dfrac{3}{5} = 2 + \dfrac{3}{5} = 2\dfrac{3}{5}$

1. $\dfrac{13}{5} = 2\dfrac{3}{5}$ 2. $\dfrac{12}{5} = 2\dfrac{2}{5}$ 3. $\dfrac{14}{11} = 1\dfrac{3}{11}$

4. $\dfrac{26}{7} = 3\dfrac{5}{7}$ 5. $\dfrac{23}{7} = 3\dfrac{2}{7}$ 6. $\dfrac{28}{5} = 5\dfrac{3}{5}$

7. $\dfrac{37}{11} = 3\dfrac{4}{11}$ 8. $\dfrac{38}{8} = 4\dfrac{3}{4}$ 9. $\dfrac{16}{7} = 2\dfrac{2}{7}$

10. $\dfrac{24}{5} = 4\dfrac{4}{5}$ 11. $\dfrac{27}{10} = 2\dfrac{7}{10}$ 12. $\dfrac{22}{12} = 1\dfrac{5}{6}$

13. $\dfrac{40}{7} = 5\dfrac{5}{7}$ 14. $\dfrac{32}{11} = 2\dfrac{10}{11}$ 15. $\dfrac{25}{7} = 3\dfrac{4}{7}$

16. $\dfrac{22}{8} = 2\dfrac{3}{4}$ 17. $\dfrac{33}{7} = 4\dfrac{5}{7}$ 18. $\dfrac{28}{8} = 3\dfrac{1}{2}$

19. $\dfrac{45}{10} = 4\dfrac{1}{2}$ 20. $\dfrac{41}{9} = 4\dfrac{5}{9}$ 21. $\dfrac{33}{5} = 6\dfrac{3}{5}$

22. $\dfrac{33}{14} = 2\dfrac{5}{14}$ 23. $\dfrac{42}{13} = 3\dfrac{3}{13}$ 24. $\dfrac{47}{11} = 4\dfrac{3}{11}$

161

Change the following improper fractions to mixed numbers.

Example: $\dfrac{13}{5} = \dfrac{10}{5} + \dfrac{3}{5} = 2 + \dfrac{3}{5} = 2\dfrac{3}{5}$

1. $\dfrac{13}{2} = 6\dfrac{1}{2}$ 2. $\dfrac{47}{6} = 7\dfrac{5}{6}$ 3. $\dfrac{7}{3} = 2\dfrac{1}{3}$

4. $\dfrac{39}{7} = 5\dfrac{4}{7}$ 5. $\dfrac{67}{8} = 8\dfrac{3}{8}$ 6. $\dfrac{24}{9} = 2\dfrac{2}{3}$

7. $\dfrac{23}{4} = 5\dfrac{3}{4}$ 8. $\dfrac{75}{10} = 7\dfrac{1}{2}$ 9. $\dfrac{60}{8} = 7\dfrac{1}{2}$

10. $\dfrac{20}{7} = 2\dfrac{6}{7}$ 11. $\dfrac{46}{5} = 9\dfrac{1}{5}$ 12. $\dfrac{42}{5} = 8\dfrac{2}{5}$

13. $\dfrac{27}{8} = 3\dfrac{3}{8}$ 14. $\dfrac{86}{9} = 9\dfrac{5}{9}$ 15. $\dfrac{33}{7} = 4\dfrac{5}{7}$

16. $\dfrac{9}{6} = 1\dfrac{1}{2}$ 17. $\dfrac{49}{8} = 6\dfrac{1}{8}$ 18. $\dfrac{38}{9} = 4\dfrac{2}{9}$

19. $\dfrac{60}{9} = 6\dfrac{2}{3}$ 20. $\dfrac{19}{7} = 2\dfrac{5}{7}$ 21. $\dfrac{38}{5} = 7\dfrac{3}{5}$

22. $\dfrac{37}{5} = 7\dfrac{2}{5}$ 23. $\dfrac{49}{15} = 3\dfrac{4}{15}$ 24. $\dfrac{65}{8} = 8\dfrac{1}{8}$

162

Change the following improper fractions to mixed numbers.

Example: $\dfrac{13}{5} = \dfrac{10}{5} + \dfrac{3}{5} = 2 + \dfrac{3}{5} = 2\dfrac{3}{5}$

1. $\dfrac{44}{5} = 8\dfrac{4}{5}$ 2. $\dfrac{23}{6} = 3\dfrac{5}{6}$ 3. $\dfrac{42}{8} = 5\dfrac{1}{4}$

4. $\dfrac{18}{8} = 2\dfrac{1}{4}$ 5. $\dfrac{24}{5} = 4\dfrac{4}{5}$ 6. $\dfrac{94}{10} = 9\dfrac{4}{10}$

7. $\dfrac{47}{6} = 7\dfrac{5}{6}$ 8. $\dfrac{70}{6} = 11\dfrac{2}{3}$ 9. $\dfrac{17}{3} = 5\dfrac{2}{3}$

10. $\dfrac{66}{8} = 8\dfrac{1}{4}$ 11. $\dfrac{84}{9} = 9\dfrac{1}{3}$ 12. $\dfrac{30}{7} = 4\dfrac{2}{7}$

13. $\dfrac{36}{5} = 7\dfrac{1}{5}$ 14. $\dfrac{76}{9} = 8\dfrac{4}{9}$ 15. $\dfrac{59}{4} = 14\dfrac{3}{4}$

16. $\dfrac{22}{8} = 2\dfrac{1}{2}$ 17. $\dfrac{54}{7} = 7\dfrac{5}{7}$ 18. $\dfrac{28}{9} = 3\dfrac{1}{9}$

19. $\dfrac{40}{9} = 4\dfrac{4}{9}$ 20. $\dfrac{32}{11} = 2\dfrac{10}{11}$ 21. $\dfrac{53}{6} = 8\dfrac{5}{6}$

22. $\dfrac{47}{7} = 6\dfrac{5}{7}$ 23. $\dfrac{89}{12} = 7\dfrac{5}{12}$ 24. $\dfrac{64}{12} = 5\dfrac{1}{3}$

163

Change the following improper fractions to mixed numbers.

Example: $\dfrac{13}{5} = \dfrac{10}{5} + \dfrac{3}{5} = 2 + \dfrac{3}{5} = 2\dfrac{3}{5}$

1. $\dfrac{27}{4} = 6\dfrac{3}{4}$ 2. $\dfrac{59}{9} = 6\dfrac{5}{9}$ 3. $\dfrac{73}{8} = 9\dfrac{1}{8}$

4. $\dfrac{89}{8} = 11\dfrac{1}{8}$ 5. $\dfrac{27}{4} = 6\dfrac{3}{4}$ 6. $\dfrac{61}{5} = 12\dfrac{1}{5}$

7. $\dfrac{54}{5} = 10\dfrac{4}{5}$ 8. $\dfrac{40}{7} = 5\dfrac{5}{7}$ 9. $\dfrac{110}{12} = 9\dfrac{1}{6}$

10. $\dfrac{75}{8} = 9\dfrac{3}{8}$ 11. $\dfrac{22}{8} = 2\dfrac{3}{4}$ 12. $\dfrac{69}{9} = 7\dfrac{2}{3}$

13. $\dfrac{26}{3} = 8\dfrac{2}{3}$ 14. $\dfrac{30}{9} = 3\dfrac{1}{3}$ 15. $\dfrac{47}{8} = 5\dfrac{7}{8}$

16. $\dfrac{47}{8} = 5\dfrac{7}{8}$ 17. $\dfrac{57}{8} = 7\dfrac{1}{8}$ 18. $\dfrac{102}{11} = 9\dfrac{3}{11}$

19. $\dfrac{55}{4} = 13\dfrac{3}{4}$ 20. $\dfrac{33}{5} = 6\dfrac{3}{5}$ 21. $\dfrac{54}{7} = 7\dfrac{5}{7}$

22. $\dfrac{80}{11} = 7\dfrac{3}{11}$ 23. $\dfrac{95}{13} = 7\dfrac{4}{13}$ 24. $\dfrac{123}{15} = 8\dfrac{1}{5}$

164

Lesson 4-7   Changing improper fractions to mixed numbers

Change the following improper fractions to mixed numbers.

Example: $\dfrac{13}{5} = \dfrac{10}{5} + \dfrac{3}{5} = 2 + \dfrac{3}{5} = 2\dfrac{3}{5}$

1. $\dfrac{48}{5} = 9\dfrac{3}{5}$ 2. $\dfrac{46}{8} = 5\dfrac{3}{4}$ 3. $\dfrac{69}{7} = 9\dfrac{6}{7}$

4. $\dfrac{23}{7} = 3\dfrac{2}{7}$ 5. $\dfrac{33}{6} = 5\dfrac{1}{2}$ 6. $\dfrac{42}{5} = 8\dfrac{2}{5}$

7. $\dfrac{38}{6} = 6\dfrac{1}{3}$ 8. $\dfrac{97}{9} = 10\dfrac{7}{9}$ 9. $\dfrac{44}{9} = 4\dfrac{8}{9}$

10. $\dfrac{64}{7} = 9\dfrac{1}{7}$ 11. $\dfrac{66}{8} = 8\dfrac{1}{4}$ 12. $\dfrac{73}{6} = 12\dfrac{1}{6}$

13. $\dfrac{30}{4} = 7\dfrac{2}{7}$ 14. $\dfrac{39}{7} = 5\dfrac{4}{7}$ 15. $\dfrac{113}{9} = 12\dfrac{5}{9}$

16. $\dfrac{91}{5} = 18\dfrac{1}{5}$ 17. $\dfrac{119}{10} = 11\dfrac{9}{10}$ 18. $\dfrac{147}{12} = 12\dfrac{1}{4}$

19. $\dfrac{123}{11} = 11\dfrac{2}{11}$ 20. $\dfrac{97}{13} = 7\dfrac{6}{13}$ 21. $\dfrac{139}{15} = 9\dfrac{4}{15}$

22. $\dfrac{107}{15} = 7\dfrac{2}{15}$ 23. $\dfrac{68}{5} = 13\dfrac{3}{5}$ 24. $\dfrac{198}{20} = 9\dfrac{9}{10}$

Lesson 4-8   Changing improper fractions to mixed numbers

Change the following improper fractions to mixed numbers.

Example: $\dfrac{13}{5} = \dfrac{10}{5} + \dfrac{3}{5} = 2 + \dfrac{3}{5} = 2\dfrac{3}{5}$

1. $\dfrac{43}{4} = 10\dfrac{3}{4}$ 2. $\dfrac{54}{8} = 6\dfrac{3}{4}$ 3. $\dfrac{26}{4} = 6\dfrac{1}{2}$

4. $\dfrac{54}{7} = 7\dfrac{5}{7}$ 5. $\dfrac{75}{6} = 12\dfrac{1}{2}$ 6. $\dfrac{39}{11} = 3\dfrac{6}{11}$

7. $\dfrac{37}{3} = 12\dfrac{1}{3}$ 8. $\dfrac{80}{9} = 8\dfrac{8}{9}$ 9. $\dfrac{64}{9} = 7\dfrac{1}{9}$

10. $\dfrac{87}{9} = 9\dfrac{2}{3}$ 11. $\dfrac{77}{8} = 9\dfrac{5}{8}$ 12. $\dfrac{27}{2} = 13\dfrac{1}{2}$

13. $\dfrac{39}{2} = 19\dfrac{1}{2}$ 14. $\dfrac{87}{6} = 14\dfrac{1}{2}$ 15. $\dfrac{93}{8} = 11\dfrac{5}{8}$

16. $\dfrac{109}{7} = 15\dfrac{4}{7}$ 17. $\dfrac{97}{10} = 9\dfrac{7}{10}$ 18. $\dfrac{72}{14} = 5\dfrac{1}{7}$

19. $\dfrac{107}{13} = 8\dfrac{3}{13}$ 20. $\dfrac{100}{16} = 6\dfrac{1}{4}$ 21. $\dfrac{93}{15} = 6\dfrac{1}{5}$

22. $\dfrac{79}{17} = 4\dfrac{11}{17}$ 23. $\dfrac{79}{18} = 4\dfrac{7}{18}$ 24. $\dfrac{100}{19} = 5\dfrac{5}{19}$

Lesson 4-9   Changing improper fractions to mixed numbers

Change the following improper fractions to mixed numbers.

Example: $\dfrac{13}{5} = \dfrac{10}{5} + \dfrac{3}{5} = 2 + \dfrac{3}{5} = 2\dfrac{3}{5}$

1. $\dfrac{77}{8} = 9\dfrac{5}{8}$ 2. $\dfrac{48}{9} = 5\dfrac{1}{3}$ 3. $\dfrac{60}{7} = 8\dfrac{4}{7}$

4. $\dfrac{38}{3} = 12\dfrac{2}{3}$ 5. $\dfrac{41}{3} = 13\dfrac{2}{3}$ 6. $\dfrac{50}{3} = 16\dfrac{2}{3}$

7. $\dfrac{52}{8} = 6\dfrac{1}{2}$ 8. $\dfrac{100}{7} = 14\dfrac{2}{7}$ 9. $\dfrac{67}{9} = 7\dfrac{4}{9}$

10. $\dfrac{53}{3} = 17\dfrac{2}{3}$ 11. $\dfrac{53}{6} = 8\dfrac{5}{6}$ 12. $\dfrac{51}{4} = 12\dfrac{3}{4}$

13. $\dfrac{45}{7} = 6\dfrac{3}{7}$ 14. $\dfrac{35}{4} = 8\dfrac{3}{4}$ 15. $\dfrac{90}{8} = 11\dfrac{1}{4}$

16. $\dfrac{96}{11} = 8\dfrac{8}{11}$ 17. $\dfrac{120}{13} = 9\dfrac{3}{13}$ 18. $\dfrac{115}{14} = 8\dfrac{3}{14}$

19. $\dfrac{121}{17} = 7\dfrac{2}{17}$ 20. $\dfrac{126}{19} = 6\dfrac{12}{19}$ 21. $\dfrac{127}{16} = 7\dfrac{15}{16}$

22. $\dfrac{128}{20} = 6\dfrac{2}{5}$ 23. $\dfrac{134}{12} = 11\dfrac{1}{6}$ 24. $\dfrac{155}{18} = 8\dfrac{11}{18}$

Lesson 4-10   Changing improper fractions to mixed numbers

Change the following improper fractions to mixed numbers.

Example: $\dfrac{13}{5} = \dfrac{10}{5} + \dfrac{3}{5} = 2 + \dfrac{3}{5} = 2\dfrac{3}{5}$

1. $\dfrac{75}{7} = 10\dfrac{5}{7}$ 2. $\dfrac{42}{4} = 10\dfrac{1}{2}$ 3. $\dfrac{73}{4} = 18\dfrac{1}{4}$

4. $\dfrac{30}{4} = 7\dfrac{1}{2}$ 5. $\dfrac{55}{8} = 6\dfrac{7}{8}$ 6. $\dfrac{44}{3} = 14\dfrac{2}{3}$

7. $\dfrac{47}{5} = 9\dfrac{2}{5}$ 8. $\dfrac{36}{5} = 7\dfrac{1}{5}$ 9. $\dfrac{75}{9} = 8\dfrac{1}{3}$

10. $\dfrac{91}{8} = 11\dfrac{3}{8}$ 11. $\dfrac{100}{8} = 12\dfrac{1}{2}$ 12. $\dfrac{119}{20} = 5\dfrac{19}{20}$

13. $\dfrac{89}{7} = 12\dfrac{5}{7}$ 14. $\dfrac{69}{7} = 9\dfrac{6}{7}$ 15. $\dfrac{41}{7} = 5\dfrac{6}{7}$

16. $\dfrac{187}{20} = 9\dfrac{7}{20}$ 17. $\dfrac{145}{11} = 13\dfrac{2}{11}$ 18. $\dfrac{71}{3} = 23\dfrac{2}{3}$

19. $\dfrac{140}{15} = 9\dfrac{1}{3}$ 20. $\dfrac{175}{19} = 9\dfrac{4}{19}$ 21. $\dfrac{151}{12} = 12\dfrac{7}{12}$

22. $\dfrac{155}{17} = 9\dfrac{2}{17}$ 23. $\dfrac{110}{17} = 6\dfrac{8}{17}$ 24. $\dfrac{130}{16} = 8\dfrac{1}{8}$

Lesson 5-1 Adding fractions with the same denominators

Add fractions and write the answers in simplest form.

1. $\frac{5}{7} + \frac{6}{7} = \frac{11}{7} = 1\frac{4}{7}$
2. $\frac{4}{9} + \frac{10}{9} = \frac{14}{9} = 1\frac{5}{9}$
3. $\frac{4}{5} + \frac{7}{5} = \frac{11}{5} = 2\frac{1}{5}$
4. $\frac{9}{7} + \frac{6}{7} = \frac{15}{7} = 2\frac{1}{7}$
5. $\frac{6}{10} + \frac{7}{10} = \frac{13}{10} = 1\frac{3}{10}$
6. $\frac{3}{5} + \frac{8}{5} = \frac{11}{5} = 2\frac{1}{5}$
7. $\frac{8}{13} + \frac{7}{13} = \frac{15}{13} = 1\frac{2}{13}$
8. $\frac{12}{10} + \frac{13}{10} = \frac{25}{10} = 2\frac{1}{2}$
9. $\frac{10}{9} + \frac{5}{9} = \frac{15}{9} = 1\frac{2}{3}$
10. $\frac{7}{4} + \frac{10}{4} = \frac{17}{4} = 4\frac{1}{4}$
11. $\frac{7}{6} + \frac{8}{6} = \frac{15}{6} = 2\frac{1}{2}$
12. $\frac{3}{8} + \frac{11}{8} = \frac{14}{8} = 1\frac{3}{4}$
13. $\frac{9}{8} + \frac{10}{8} = \frac{19}{8} = 2\frac{3}{8}$
14. $\frac{15}{13} + \frac{12}{13} = \frac{27}{13} = 2\frac{1}{13}$
15. $\frac{5}{4} + \frac{6}{4} = \frac{11}{4} = 2\frac{3}{4}$
16. $\frac{6}{11} + \frac{7}{11} = \frac{13}{11} = 1\frac{2}{11}$
17. $\frac{13}{11} + \frac{12}{11} = \frac{25}{11} = 2\frac{3}{11}$
18. $\frac{9}{6} + \frac{7}{6} = \frac{16}{6} = 2\frac{2}{3}$

Lesson 5-2 Adding fractions with the same denominators

Add fractions and write the answers in simplest form.

1. $\frac{4}{5} + \frac{7}{5} = \frac{11}{5} = 2\frac{1}{5}$
2. $\frac{9}{5} + \frac{7}{5} = \frac{16}{5} = 3\frac{1}{5}$
3. $\frac{7}{8} + \frac{6}{8} = \frac{13}{8} = 1\frac{5}{8}$
4. $\frac{6}{7} + \frac{9}{7} = \frac{15}{7} = 2\frac{1}{7}$
5. $\frac{5}{4} + \frac{6}{4} = \frac{11}{4} = 2\frac{3}{4}$
6. $\frac{5}{4} + \frac{9}{4} = \frac{14}{4} = 3\frac{1}{2}$
7. $\frac{7}{3} + \frac{4}{3} = \frac{10}{3} = 3\frac{1}{3}$
8. $\frac{9}{10} + \frac{7}{10} = \frac{16}{10} = 1\frac{3}{5}$
9. $\frac{5}{6} + \frac{9}{6} = \frac{14}{6} = 2\frac{1}{3}$
10. $\frac{11}{9} + \frac{15}{9} = \frac{26}{9} = 2\frac{8}{9}$
11. $\frac{13}{10} + \frac{23}{10} = \frac{36}{10} = 3\frac{3}{5}$
12. $\frac{13}{8} + \frac{5}{8} = \frac{18}{8} = 2\frac{1}{4}$
13. $\frac{9}{20} + \frac{13}{20} = \frac{22}{20} = 1\frac{1}{10}$
14. $\frac{5}{2} + \frac{8}{2} = \frac{13}{2} = 6\frac{1}{2}$
15. $\frac{11}{13} + \frac{14}{13} = \frac{25}{13} = 1\frac{12}{13}$
16. $\frac{13}{3} + \frac{12}{3} = \frac{25}{3} = 8\frac{1}{3}$
17. $\frac{17}{15} + \frac{12}{15} = \frac{29}{15} = 1\frac{14}{15}$
18. $\frac{14}{17} + \frac{15}{17} = \frac{29}{17} = 1\frac{12}{17}$

Lesson 5-3 Adding fractions with the same denominators

Add fractions and write the answers in simplest form.

1. $1\frac{2}{5} + \frac{6}{5} = 1\frac{8}{5} = 2\frac{3}{5}$
2. $2\frac{2}{3} + \frac{5}{3} = 2\frac{7}{3} = 4\frac{1}{3}$
3. $2\frac{5}{7} + 1\frac{8}{7} = 3\frac{13}{7} = 4\frac{6}{7}$
4. $1\frac{4}{5} + 1\frac{7}{5} = 2\frac{11}{5} = 4\frac{1}{5}$
5. $\frac{8}{3} + 1\frac{2}{3} = 1\frac{10}{3} = 4\frac{1}{3}$
6. $1\frac{5}{9} + 1\frac{10}{9} = 2\frac{15}{9} = 3\frac{2}{3}$
7. $1\frac{7}{6} + 1\frac{8}{6} = 2\frac{15}{6} = 4\frac{1}{2}$
8. $2\frac{7}{6} + \frac{3}{6} = 2\frac{10}{6} = 3\frac{2}{3}$
9. $2\frac{3}{8} + 2\frac{9}{8} = 4\frac{12}{8} = 5\frac{1}{2}$
10. $2\frac{3}{10} + 1\frac{13}{10} = 3\frac{16}{10} = 4\frac{3}{5}$
11. $1\frac{5}{9} + 1\frac{6}{9} = 2\frac{11}{9} = 3\frac{2}{9}$
12. $\frac{9}{8} + 2\frac{5}{8} = 2\frac{14}{8} = 3\frac{3}{4}$
13. $\frac{12}{10} + 2\frac{13}{10} = 2\frac{25}{10} = 4\frac{1}{2}$
14. $1\frac{8}{7} + 2\frac{9}{7} = 3\frac{17}{7} = 5\frac{3}{7}$
15. $\frac{15}{11} + 1\frac{13}{11} = 1\frac{18}{11} = 2\frac{7}{11}$
16. $\frac{5}{12} + 1\frac{11}{12} = 1\frac{16}{12} = 2\frac{1}{3}$
17. $2\frac{5}{12} + 1\frac{13}{12} = 3\frac{18}{12} = 4\frac{1}{2}$
18. $2\frac{3}{15} + 2\frac{4}{15} = 4\frac{7}{15}$

Lesson 5-4 Adding fractions with the same denominators

Add fractions and write the answers in simplest form.

1. $2\frac{5}{3} + 1\frac{5}{3} = 3\frac{10}{3} = 6\frac{1}{3}$
2. $2\frac{3}{4} + 1\frac{7}{4} = 3\frac{10}{4} = 5\frac{1}{2}$
3. $1\frac{3}{7} + 2\frac{8}{7} = 3\frac{11}{7} = 4\frac{4}{7}$
4. $3\frac{4}{10} + 3\frac{7}{10} = 6\frac{11}{10} = 7\frac{1}{10}$
5. $1\frac{12}{11} + \frac{14}{11} = 1\frac{26}{11} = 3\frac{4}{11}$
6. $2\frac{5}{9} + 1\frac{10}{9} = 3\frac{15}{9} = 4\frac{2}{3}$
7. $3\frac{3}{5} + 2\frac{9}{5} = 5\frac{12}{5} = 7\frac{2}{5}$
8. $\frac{4}{7} + 3\frac{6}{7} = 3\frac{10}{7} = 4\frac{3}{7}$
9. $\frac{15}{12} + 3\frac{7}{12} = 3\frac{22}{12} = 4\frac{5}{6}$
10. $3\frac{3}{8} + \frac{7}{8} = 3\frac{10}{8} = 4\frac{1}{4}$
11. $1\frac{21}{20} + 3\frac{22}{20} = 4\frac{43}{20} = 6\frac{3}{20}$
12. $1\frac{9}{20} + 2\frac{14}{20} = 3\frac{23}{20} = 4\frac{3}{20}$
13. $2\frac{11}{18} + 3\frac{10}{18} = 5\frac{21}{18} = 6\frac{1}{6}$
14. $1\frac{13}{15} + 1\frac{18}{15} = 2\frac{31}{15} = 4\frac{1}{15}$
15. $\frac{8}{15} + 3\frac{11}{15} = 3\frac{19}{15} = 4\frac{4}{15}$
16. $3\frac{12}{13} + 1\frac{12}{13} = 4\frac{24}{13} = 5\frac{11}{13}$
17. $3\frac{4}{9} + 3\frac{5}{9} = 6\frac{9}{9} = 7$
18. $2\frac{8}{19} + 3\frac{7}{19} = 5\frac{15}{19}$

Lesson 5-5   Adding fractions with the same denominators

Add fractions and write the answers in simplest form.

1. $3\frac{3}{5} + 2\frac{4}{5} = 5\frac{7}{5} = 6\frac{2}{5}$ 2. $4\frac{7}{6} + 1\frac{4}{6} = 5\frac{10}{6} = 6\frac{2}{3}$

3. $1\frac{7}{10} + 3\frac{8}{10} = 4\frac{15}{10} = 5\frac{1}{2}$ 4. $3\frac{11}{15} + 2\frac{6}{15} = 5\frac{17}{15} = 6\frac{2}{15}$

5. $4\frac{10}{9} + 2\frac{9}{9} = 6\frac{19}{9} = 8\frac{1}{9}$ 6. $4\frac{10}{9} + 3\frac{11}{9} = 7\frac{21}{9} = 9\frac{1}{3}$

7. $3\frac{5}{7} + 3\frac{6}{7} = 6\frac{11}{7} = 7\frac{4}{7}$ 8. $2\frac{7}{17} + 3\frac{18}{17} = 5\frac{25}{17} = 6\frac{8}{17}$

9. $2\frac{5}{12} + 3\frac{10}{12} = 5\frac{15}{12} = 6\frac{1}{4}$ 10. $4\frac{5}{8} + 2\frac{9}{8} = 6\frac{14}{8} = 7\frac{3}{4}$

11. $1\frac{11}{25} + 4\frac{12}{25} = 5\frac{23}{25}$ 12. $1\frac{9}{10} + 4\frac{9}{10} = 5\frac{18}{10} = 6\frac{4}{5}$

13. $3\frac{9}{8} + 4\frac{11}{8} = 7\frac{20}{8} = 9\frac{1}{2}$ 14. $\frac{7}{11} + \frac{9}{11} = \frac{16}{11} = 1\frac{5}{11}$

15. $4\frac{13}{16} + 2\frac{15}{16} = 6\frac{28}{16} = 7\frac{3}{4}$ 16. $4\frac{9}{16} + 4\frac{10}{16} = 8\frac{19}{16} = 9\frac{3}{16}$

17. $4\frac{17}{21} + 4\frac{18}{21} = 8\frac{35}{21} = 9\frac{2}{3}$ 18. $3\frac{21}{23} + 4\frac{22}{23} = 7\frac{43}{23} = 8\frac{20}{23}$

Lesson 5-6   Adding fractions with the same denominators

Add fractions and write the answers in simplest form.

1. $3\frac{5}{7} + 1\frac{4}{7} = 4\frac{9}{7} = 5\frac{2}{7}$ 2. $2\frac{7}{9} + 2\frac{10}{9} = 4\frac{17}{9} = 5\frac{8}{9}$

3. $4\frac{4}{9} + 2\frac{10}{9} = 6\frac{13}{9} = 7\frac{4}{9}$ 4. $4\frac{19}{20} + 1\frac{11}{20} = 5\frac{30}{20} = 6\frac{1}{2}$

5. $3\frac{11}{14} + 4\frac{13}{14} = 7\frac{24}{14} = 8\frac{5}{7}$ 6. $2\frac{5}{7} + 3\frac{8}{7} = 5\frac{13}{7} = 6\frac{6}{7}$

7. $3\frac{3}{4} + 2\frac{6}{4} = 5\frac{9}{4} = 7\frac{1}{4}$ 8. $3\frac{13}{10} + \frac{11}{10} = 3\frac{24}{10} = 5\frac{2}{5}$

9. $2\frac{7}{11} + 3\frac{9}{11} = 5\frac{16}{11} = 6\frac{5}{11}$ 10. $1\frac{13}{16} + 4\frac{9}{16} = 5\frac{22}{16} = 6\frac{3}{8}$

11. $4\frac{18}{17} + 3\frac{11}{17} = 7\frac{29}{17} = 8\frac{12}{17}$ 12. $3\frac{7}{11} + 2\frac{9}{11} = 5\frac{16}{11} = 6\frac{5}{11}$

13. $1\frac{21}{25} + 3\frac{17}{25} = 4\frac{38}{25} = 5\frac{13}{25}$ 14. $2\frac{10}{12} + 4\frac{13}{12} = 6\frac{23}{12} = 7\frac{11}{12}$

15. $\frac{10}{21} + 2\frac{11}{21} = 2\frac{21}{21} = 3$ 16. $4\frac{19}{24} + 3\frac{15}{24} = 7\frac{34}{24} = 8\frac{5}{12}$

17. $2\frac{4}{15} + 4\frac{13}{15} = 6\frac{28}{15} = 7\frac{13}{15}$ 18. $4\frac{12}{18} + 2\frac{20}{18} = 6\frac{32}{18} = 7\frac{7}{9}$

Lesson 5-7   Adding fractions with the same denominators

Add fractions and write the answers in simplest form.

1. $3\frac{6}{5} + 3\frac{3}{5} = 6\frac{9}{5} = 7\frac{4}{5}$ 2. $1\frac{6}{7} + 4\frac{9}{7} = 5\frac{15}{7} = 7\frac{1}{7}$

3. $4\frac{12}{16} + 2\frac{11}{16} = 6\frac{23}{16} = 7\frac{7}{16}$ 4. $4\frac{5}{8} + 2\frac{7}{8} = 6\frac{12}{8} = 7\frac{1}{2}$

5. $4\frac{7}{15} + 3\frac{8}{15} = 7\frac{15}{15} = 8$ 6. $5\frac{17}{20} + 3\frac{7}{20} = 8\frac{24}{20} = 9\frac{1}{5}$

7. $5\frac{11}{8} + 2\frac{10}{8} = 7\frac{21}{8} = 9\frac{5}{8}$ 8. $4\frac{11}{14} + 1\frac{13}{14} = 5\frac{24}{14} = 6\frac{5}{7}$

9. $2\frac{8}{9} + 4\frac{8}{9} = 6\frac{16}{9} = 7\frac{7}{9}$ 10. $2\frac{21}{23} + 4\frac{12}{23} = 6\frac{33}{23} = 7\frac{10}{23}$

11. $3\frac{10}{24} + \frac{17}{24} = 6\frac{27}{24} = 7\frac{1}{8}$ 12. $4\frac{9}{15} + 4\frac{10}{15} = 8\frac{19}{15} = 9\frac{4}{15}$

13. $3\frac{8}{11} + 4\frac{10}{11} = 7\frac{18}{11} = 8\frac{7}{11}$ 14. $3\frac{10}{12} + 5\frac{6}{12} = 8\frac{16}{12} = 9\frac{1}{3}$

15. $2\frac{21}{25} + 3\frac{22}{25} = 5\frac{43}{25} = 6\frac{18}{25}$ 16. $5\frac{19}{17} + \frac{18}{17} = 5\frac{37}{17} = 7\frac{3}{17}$

17. $2\frac{12}{17} + 5\frac{13}{17} = 7\frac{25}{17} = 8\frac{8}{17}$ 18. $5\frac{15}{19} + 2\frac{14}{19} = 7\frac{29}{19} = 8\frac{10}{19}$

Lesson 5-8   Adding fractions with the same denominators

Add fractions and write the answers in simplest form.

1. $3\frac{5}{6} + 2\frac{3}{6} = 5\frac{8}{6} = 6\frac{1}{3}$ 2. $2\frac{5}{8} + 2\frac{7}{8} = 4\frac{12}{8} = 5\frac{1}{2}$

3. $1\frac{7}{14} + 4\frac{9}{14} = 5\frac{16}{14} = 6\frac{1}{7}$ 4. $3\frac{5}{4} + 4\frac{2}{4} = 7\frac{7}{4} = 8\frac{3}{4}$

5. $3\frac{11}{8} + 3\frac{7}{8} = 6\frac{18}{8} = 8\frac{1}{4}$ 6. $4\frac{9}{13} + 3\frac{11}{13} = 7\frac{20}{13} = 8\frac{7}{13}$

7. $2\frac{18}{22} + 3\frac{11}{22} = 5\frac{29}{22} = 6\frac{7}{22}$ 8. $3\frac{8}{19} + 2\frac{15}{19} = 5\frac{23}{19} = 6\frac{4}{19}$

9. $5\frac{4}{9} + \frac{10}{9} = 5\frac{14}{9} = 6\frac{5}{9}$ 10. $4\frac{15}{14} + 3\frac{11}{14} = 7\frac{26}{14} = 8\frac{6}{7}$

11. $4\frac{13}{16} + 2\frac{7}{16} = 6\frac{20}{16} = 7\frac{1}{4}$ 12. $1\frac{9}{12} + 5\frac{9}{12} = 6\frac{18}{12} = 7\frac{1}{2}$

13. $3\frac{9}{20} + 4\frac{12}{20} = 7\frac{21}{20} = 8\frac{1}{20}$ 14. $\frac{12}{15} + \frac{6}{15} = \frac{18}{15} = 1\frac{1}{5}$

15. $2\frac{14}{11} + 3\frac{13}{11} = 5\frac{27}{11} = 7\frac{5}{11}$ 16. $3\frac{8}{13} + 4\frac{7}{13} = 7\frac{15}{13} = 8\frac{2}{13}$

17. $3\frac{15}{16} + 5\frac{11}{16} = 8\frac{26}{16} = 9\frac{5}{8}$ 18. $5\frac{17}{21} + 2\frac{19}{21} = 7\frac{36}{21} = 8\frac{5}{7}$

Lesson 5-9 Adding fractions with the same denominators

Add fractions and write the answers in simplest form.

1. $5\frac{6}{5} + 1\frac{7}{5} = 6\frac{13}{5} = 8\frac{3}{5}$

2. $2\frac{9}{5} + 1\frac{13}{5} = 3\frac{22}{5} = 7\frac{2}{5}$

3. $3\frac{4}{9} + 4\frac{11}{9} = 7\frac{15}{9} = 8\frac{2}{3}$

4. $4\frac{15}{17} + 3\frac{16}{17} = 7\frac{31}{17} = 8\frac{14}{17}$

5. $4\frac{9}{11} + 4\frac{8}{11} = 8\frac{17}{11} = 9\frac{6}{11}$

6. $2\frac{5}{8} + 5\frac{9}{8} = 7\frac{14}{8} = 8\frac{3}{4}$

7. $2\frac{9}{25} + 3\frac{13}{25} = 5\frac{22}{25}$

8. $4\frac{6}{13} + 4\frac{9}{13} = 8\frac{15}{13} = 9\frac{2}{13}$

9. $1\frac{10}{13} + 3\frac{11}{13} = 4\frac{21}{13} = 5\frac{8}{13}$

10. $3\frac{20}{22} + 4\frac{21}{22} = 7\frac{41}{22} = 8\frac{19}{22}$

11. $3\frac{13}{17} + 2\frac{14}{17} = 5\frac{27}{17} = 6\frac{10}{17}$

12. $5\frac{13}{18} + 2\frac{15}{18} = 7\frac{28}{18} = 8\frac{5}{9}$

13. $4\frac{20}{12} + 3\frac{13}{12} = 7\frac{33}{12} = 9\frac{3}{4}$

14. $1\frac{20}{16} + 3\frac{17}{16} = 4\frac{37}{16} = 6\frac{5}{16}$

15. $4\frac{17}{20} + 2\frac{18}{20} = 6\frac{35}{20} = 7\frac{3}{4}$

16. $3\frac{7}{12} + 2\frac{15}{12} = 5\frac{22}{12} = 6\frac{5}{6}$

17. $3\frac{19}{23} + 3\frac{20}{23} = 6\frac{39}{23} = 7\frac{16}{23}$

18. $3\frac{7}{24} + 5\frac{17}{24} = 8\frac{24}{24} = 9$

Lesson 5-10 Adding fractions with the same denominators

Add fractions and write the answers in simplest form.

1. $2\frac{11}{5} + 5\frac{12}{5} = 7\frac{23}{5} = 11\frac{3}{5}$

2. $4\frac{7}{8} + 4\frac{7}{8} = 8\frac{14}{8} = 9\frac{3}{4}$

3. $4\frac{6}{13} + 2\frac{8}{13} = 6\frac{14}{13} = 7\frac{1}{13}$

4. $1\frac{6}{10} + 5\frac{9}{10} = 6\frac{15}{10} = 7\frac{1}{2}$

5. $3\frac{8}{7} + 5\frac{9}{7} = 8\frac{17}{7} = 10\frac{3}{7}$

6. $4\frac{10}{9} + 3\frac{12}{9} = 7\frac{22}{9} = 9\frac{4}{9}$

7. $5\frac{13}{15} + 3\frac{14}{15} = 8\frac{27}{15} = 9\frac{4}{5}$

8. $2\frac{7}{16} + 2\frac{18}{16} = 4\frac{25}{16} = 5\frac{9}{16}$

9. $5\frac{4}{25} + 2\frac{23}{25} = 7\frac{27}{25} = 8\frac{2}{25}$

10. $4\frac{19}{24} + 5\frac{17}{24} = 9\frac{36}{24} = 10\frac{1}{2}$

11. $2\frac{12}{18} + 4\frac{15}{18} = 6\frac{27}{18} = 7\frac{1}{2}$

12. $4\frac{21}{14} + 3\frac{13}{14} = 7\frac{34}{14} = 9\frac{3}{7}$

13. $5\frac{11}{17} + \frac{13}{17} = 5\frac{24}{17} = 6\frac{7}{17}$

14. $3\frac{5}{11} + 4\frac{9}{11} = 7\frac{14}{11} = 8\frac{3}{11}$

15. $2\frac{20}{23} + 3\frac{22}{23} = 5\frac{42}{23} = 6\frac{19}{23}$

16. $3\frac{19}{23} + 3\frac{21}{23} = 6\frac{40}{23} = 7\frac{17}{23}$

17. $4\frac{9}{12} + 3\frac{12}{12} = 7\frac{21}{12} = 8\frac{3}{4}$

18. $5\frac{24}{17} + 1\frac{11}{17} = 6\frac{35}{17} = 8\frac{1}{17}$

Lesson 6-1 Subtracting fractions with the same denominators

Subtract fractions and write the answers in simplest form.

1. $1\frac{2}{3} - 1\frac{1}{3} = \frac{5}{3} - \frac{1}{3} = 1\frac{1}{3}$

2. $1\frac{2}{9} - \frac{7}{9} = \frac{11}{9} - \frac{7}{9} = \frac{4}{9}$

3. $2\frac{3}{7} - \frac{5}{7} = \frac{17}{7} - \frac{5}{7} = 1\frac{5}{7}$

4. $1\frac{3}{10} - \frac{9}{10} = \frac{13}{10} - \frac{9}{10} = \frac{2}{5}$

5. $1\frac{1}{5} - \frac{4}{5} = \frac{6}{5} - \frac{4}{5} = \frac{2}{5}$

6. $1\frac{3}{5} - \frac{2}{5} = \frac{8}{5} - \frac{2}{5} = 1\frac{1}{5}$

7. $2\frac{2}{11} - 1\frac{5}{11} = \frac{23}{11} - \frac{16}{11} = \frac{7}{11}$

8. $1\frac{7}{8} - \frac{5}{8} = \frac{15}{8} - \frac{5}{8} = 1\frac{1}{4}$

9. $1\frac{2}{9} - \frac{5}{9} = \frac{11}{9} - \frac{5}{9} = \frac{2}{3}$

10. $2\frac{3}{11} - 1\frac{10}{11} = \frac{25}{11} - \frac{21}{11} = \frac{4}{11}$

11. $2\frac{3}{10} - 1\frac{1}{10} = \frac{23}{10} - \frac{11}{10} = 1\frac{1}{5}$

12. $2\frac{4}{7} - 1\frac{8}{7} = \frac{18}{7} - \frac{15}{7} = \frac{3}{7}$

13. $1\frac{7}{13} - \frac{10}{13} = \frac{20}{13} - \frac{10}{13} = \frac{10}{13}$

14. $2\frac{1}{14} - 1\frac{9}{14} = \frac{29}{14} - \frac{23}{14} = \frac{3}{7}$

15. $2\frac{1}{20} - \frac{11}{20} = \frac{41}{20} - \frac{11}{20} = 1\frac{1}{2}$

16. $2\frac{5}{6} - 1\frac{7}{6} = \frac{17}{6} - \frac{13}{6} = \frac{2}{3}$

17. $1\frac{12}{15} - \frac{14}{15} = \frac{27}{15} - \frac{14}{15} = \frac{13}{15}$

18. $2\frac{3}{17} - 1\frac{13}{17} = \frac{37}{17} - \frac{30}{17} = \frac{7}{17}$

Lesson 6-2 Subtracting fractions with the same denominators

Subtract fractions and write the answers in simplest form.

1. $2 - \frac{5}{6} = \frac{12}{6} - \frac{5}{6} = 1\frac{1}{6}$

2. $3 - \frac{9}{7} = \frac{21}{7} - \frac{9}{7} = 1\frac{5}{7}$

3. $\frac{21}{8} - \frac{14}{8} = \frac{21}{8} - \frac{14}{8} = \frac{7}{8}$

4. $2\frac{3}{5} - \frac{7}{5} = \frac{13}{5} - \frac{7}{5} = 1\frac{1}{5}$

5. $2\frac{3}{10} - 1\frac{9}{10} = \frac{23}{10} - \frac{19}{10} = \frac{2}{5}$

6. $2\frac{2}{21} - 1\frac{19}{21} = \frac{44}{21} - \frac{40}{21} = \frac{4}{21}$

7. $3\frac{1}{13} - 2\frac{9}{13} = \frac{40}{13} - \frac{35}{13} = \frac{5}{13}$

8. $\frac{24}{9} - 1\frac{2}{9} = \frac{24}{9} - \frac{11}{9} = 1\frac{4}{9}$

9. $1\frac{2}{7} - \frac{8}{7} = \frac{9}{7} - \frac{8}{7} = \frac{1}{7}$

10. $3\frac{1}{12} - 2\frac{5}{12} = \frac{37}{12} - \frac{29}{12} = \frac{2}{3}$

11. $2\frac{11}{17} - \frac{9}{17} = \frac{45}{17} - \frac{9}{17} = 2\frac{2}{17}$

12. $\frac{13}{4} - 2\frac{1}{4} = \frac{13}{4} - \frac{9}{4} = 1$

13. $1\frac{5}{11} - \frac{10}{11} = \frac{16}{11} - \frac{10}{11} = \frac{6}{11}$

14. $2\frac{5}{18} - \frac{17}{18} = \frac{41}{18} - \frac{17}{18} = 1\frac{1}{3}$

15. $3\frac{4}{25} - 1\frac{23}{25} = \frac{79}{25} - \frac{48}{25} = 1\frac{6}{25}$

16. $1\frac{7}{23} - \frac{2}{23} = \frac{30}{23} - \frac{2}{23} = 1\frac{5}{23}$

17. $2\frac{7}{19} - \frac{20}{19} = \frac{45}{19} - \frac{20}{19} = 1\frac{6}{19}$

18. $\frac{33}{16} - 1\frac{1}{16} = \frac{33}{16} - \frac{17}{16} = 1$

Lesson 6-3 Subtracting fractions with the same denominators

Subtract fractions and write the answers in simplest form.

1. $3\frac{1}{6} - 1\frac{2}{6} = \frac{19}{6} - \frac{8}{6} = 1\frac{5}{6}$

2. $3\frac{1}{8} - 1\frac{3}{8} = \frac{25}{8} - \frac{11}{8} = 1\frac{3}{4}$

3. $2\frac{1}{7} - \frac{5}{7} = \frac{15}{7} - \frac{5}{7} = 1\frac{3}{7}$

4. $2\frac{2}{9} - \frac{11}{9} = \frac{20}{9} - \frac{11}{9} = 1$

5. $4 - 2\frac{2}{3} = \frac{12}{3} - \frac{8}{3} = 1\frac{1}{3}$

6. $2\frac{3}{14} - 1\frac{11}{14} = \frac{31}{14} - \frac{25}{14} = \frac{3}{7}$

7. $1\frac{3}{14} - \frac{11}{14} = \frac{17}{14} - \frac{11}{14} = \frac{3}{7}$

8. $4\frac{1}{13} - 2\frac{9}{13} = \frac{53}{13} - \frac{35}{13} = 1\frac{5}{13}$

9. $\frac{33}{24} - 1\frac{1}{24} = \frac{33}{24} - \frac{25}{24} = \frac{1}{3}$

10. $3 - 1\frac{6}{7} = \frac{21}{7} - \frac{13}{7} = 1\frac{1}{7}$

11. $3\frac{2}{15} - 1\frac{7}{15} = \frac{47}{15} - \frac{22}{15} = 1\frac{2}{3}$

12. $2\frac{3}{11} - \frac{13}{11} = \frac{25}{11} - \frac{13}{11} = 1\frac{1}{12}$

13. $4\frac{3}{23} - 1\frac{22}{23} = \frac{95}{23} - \frac{45}{23} = 2\frac{4}{23}$

14. $4\frac{2}{27} - 3\frac{3}{27} = \frac{110}{27} - \frac{84}{27} = \frac{26}{27}$

15. $\frac{40}{18} - \frac{16}{18} = \frac{40}{18} - \frac{16}{18} = 1\frac{1}{3}$

16. $3\frac{4}{17} - 1\frac{16}{17} = \frac{55}{17} - \frac{33}{17} = 1\frac{5}{17}$

17. $4\frac{5}{12} - 2\frac{13}{12} = \frac{53}{12} - \frac{37}{12} = 1\frac{1}{3}$

18. $4\frac{3}{14} - 2\frac{15}{14} = \frac{59}{14} - \frac{43}{14} = 1\frac{1}{7}$

181

Lesson 6-4 Subtracting fractions with the same denominators

Subtract fractions and write the answers in simplest form.

1. $4\frac{1}{3} - 2\frac{2}{3} = \frac{13}{3} - \frac{8}{3} = 1\frac{2}{3}$

2. $2\frac{3}{5} - 1\frac{4}{5} = \frac{13}{5} - \frac{9}{5} = \frac{4}{5}$

3. $3\frac{1}{4} - 1\frac{3}{4} = \frac{13}{4} - \frac{7}{4} = 1\frac{1}{2}$

4. $4\frac{2}{7} - 2\frac{6}{7} = \frac{30}{7} - \frac{20}{7} = 1\frac{3}{7}$

5. $2\frac{5}{13} - \frac{9}{13} = \frac{31}{13} - \frac{9}{13} = 1\frac{9}{13}$

6. $3\frac{1}{8} - 2\frac{5}{8} = \frac{25}{8} - \frac{21}{8} = \frac{1}{2}$

7. $5 - 4\frac{5}{16} = \frac{80}{16} - \frac{69}{16} = \frac{11}{16}$

8. $4\frac{5}{14} - 1\frac{12}{14} = \frac{61}{14} - \frac{26}{14} = 2\frac{1}{2}$

9. $2\frac{4}{9} - 1\frac{8}{9} = \frac{22}{9} - \frac{17}{9} = \frac{5}{9}$

10. $3\frac{5}{21} - \frac{19}{21} = \frac{68}{21} - \frac{19}{21} = 2\frac{1}{3}$

11. $5\frac{1}{11} - 3\frac{2}{11} = \frac{56}{11} - \frac{35}{11} = 1\frac{10}{11}$

12. $5\frac{7}{20} - 2\frac{17}{20} = \frac{107}{20} - \frac{57}{20} = 2\frac{1}{2}$

13. $3\frac{15}{26} - 1\frac{21}{26} = \frac{93}{26} - \frac{47}{26} = 1\frac{10}{13}$

14. $5\frac{3}{16} - 3\frac{15}{16} = \frac{83}{16} - \frac{63}{16} = 1\frac{1}{4}$

15. $4\frac{7}{15} - 1\frac{4}{15} = \frac{67}{15} - \frac{19}{15} = 3\frac{1}{5}$

16. $4\frac{1}{12} - 2\frac{11}{12} = \frac{49}{12} - \frac{35}{12} = 1\frac{1}{6}$

17. $3\frac{3}{17} - \frac{16}{17} = \frac{54}{17} - \frac{16}{17} = 2\frac{4}{17}$

18. $5\frac{7}{18} - 2\frac{13}{18} = \frac{97}{18} - \frac{49}{18} = 2\frac{2}{3}$

182

Lesson 6-5 Subtracting fractions with the same denominators

Subtract fractions and write the answers in simplest form.

1. $6\frac{1}{4} - 2\frac{3}{4} = \frac{25}{4} - \frac{11}{4} = 3\frac{1}{2}$

2. $4\frac{2}{9} - 2\frac{7}{9} = \frac{38}{9} - \frac{25}{9} = 1\frac{4}{9}$

3. $4\frac{3}{8} - 3\frac{7}{8} = \frac{35}{8} - \frac{31}{8} = \frac{1}{2}$

4. $5\frac{1}{5} - 1\frac{4}{5} = \frac{26}{5} - \frac{9}{5} = 3\frac{2}{5}$

5. $5\frac{2}{21} - 3\frac{17}{21} = \frac{107}{21} - \frac{80}{21} = 1\frac{2}{7}$

6. $3\frac{3}{17} - \frac{15}{17} = \frac{54}{17} - \frac{15}{17} = 2\frac{5}{17}$

7. $6\frac{3}{17} - 3\frac{9}{17} = \frac{105}{17} - \frac{60}{17} = 2\frac{11}{17}$

8. $5\frac{9}{13} - 2\frac{11}{13} = \frac{74}{13} - \frac{37}{13} = 2\frac{11}{13}$

9. $4\frac{11}{25} - 1\frac{23}{25} = \frac{111}{25} - \frac{48}{25} = 2\frac{13}{25}$

10. $6\frac{5}{18} - 3\frac{17}{18} = \frac{113}{18} - \frac{71}{18} = 2\frac{1}{3}$

11. $3\frac{5}{13} - 2\frac{1}{13} = \frac{44}{13} - \frac{27}{13} = 1\frac{4}{13}$

12. $6\frac{3}{16} - 2\frac{13}{16} = \frac{99}{16} - \frac{45}{16} = 3\frac{3}{8}$

13. $6\frac{7}{30} - 5\frac{23}{30} = \frac{187}{30} - \frac{173}{30} = \frac{7}{15}$

14. $5 - 3\frac{15}{19} = \frac{95}{19} - \frac{72}{19} = 1\frac{4}{19}$

15. $5\frac{2}{26} - 1\frac{19}{26} = \frac{132}{26} - \frac{47}{26} = 3\frac{7}{26}$

16. $4\frac{4}{18} - 1\frac{15}{18} = \frac{76}{18} - \frac{33}{18} = 2\frac{7}{18}$

17. $6 - \frac{3}{17} = \frac{102}{17} - \frac{3}{17} = 5\frac{14}{17}$

18. $3\frac{2}{23} - \frac{22}{23} = \frac{71}{23} - \frac{22}{23} = 2\frac{3}{23}$

183

Lesson 6-6 Subtracting fractions with the same denominators

Subtract fractions and write the answers in simplest form.

1. $6 - \frac{7}{3} = \frac{18}{3} - \frac{7}{3} = 3\frac{2}{3}$

2. $5 - \frac{13}{8} = \frac{40}{8} - \frac{13}{8} = 3\frac{3}{8}$

3. $4\frac{1}{4} - 2\frac{2}{4} = \frac{17}{4} - \frac{10}{4} = 1\frac{3}{4}$

4. $2\frac{2}{7} - \frac{5}{7} = \frac{16}{7} - \frac{5}{7} = 1\frac{4}{7}$

5. $5\frac{3}{9} - 2\frac{8}{9} = \frac{48}{9} - \frac{26}{9} = 2\frac{4}{9}$

6. $5\frac{8}{13} - 1\frac{10}{13} = \frac{73}{13} - \frac{23}{13} = 3\frac{11}{13}$

7. $3\frac{1}{6} - \frac{5}{6} = \frac{19}{6} - \frac{5}{6} = 2\frac{1}{3}$

8. $4\frac{6}{23} - 3\frac{19}{23} = \frac{98}{23} - \frac{88}{23} = \frac{10}{23}$

9. $6\frac{7}{13} - 3\frac{12}{13} = \frac{85}{13} - \frac{51}{13} = 2\frac{8}{13}$

10. $5\frac{7}{15} - 2\frac{13}{15} = \frac{82}{15} - \frac{43}{15} = 2\frac{3}{5}$

11. $4\frac{2}{14} - 1\frac{5}{14} = \frac{58}{14} - \frac{19}{14} = 2\frac{11}{14}$

12. $6\frac{1}{20} - 5\frac{9}{20} = \frac{121}{20} - \frac{109}{20} = \frac{3}{5}$

13. $6\frac{3}{19} - 4\frac{11}{19} = \frac{117}{19} - \frac{87}{19} = 1\frac{11}{19}$

14. $4\frac{5}{17} - 2\frac{15}{17} = \frac{73}{17} - \frac{49}{17} = 1\frac{7}{17}$

15. $5\frac{6}{17} - 1\frac{13}{17} = \frac{91}{17} - \frac{30}{17} = 3\frac{10}{17}$

16. $6\frac{3}{28} - 1\frac{25}{28} = \frac{171}{28} - \frac{53}{28} = 4\frac{3}{14}$

17. $6\frac{1}{11} - 2\frac{8}{11} = \frac{67}{11} - \frac{30}{11} = 3\frac{4}{11}$

18. $6\frac{5}{18} - 4\frac{16}{18} = \frac{113}{18} - \frac{88}{18} = 1\frac{7}{18}$

184

Lesson 6-7 Subtracting fractions with the same denominators

Subtract fractions and write the answers in simplest form.

1. $1\frac{1}{6} - \frac{5}{6} = \frac{7}{6} - \frac{5}{6} = \frac{1}{3}$

2. $3\frac{1}{8} - 2\frac{7}{8} = \frac{25}{8} - \frac{23}{8} = \frac{1}{4}$

3. $2\frac{3}{7} - 1\frac{6}{7} = \frac{17}{7} - \frac{13}{7} = \frac{4}{7}$

4. $4\frac{2}{9} - \frac{8}{9} = \frac{38}{9} - \frac{8}{9} = 3\frac{1}{3}$

5. $4\frac{2}{15} - 2\frac{11}{15} = \frac{62}{15} - \frac{41}{15} = 1\frac{2}{5}$

6. $2\frac{3}{13} - \frac{12}{13} = \frac{29}{13} - \frac{12}{13} = 1\frac{4}{13}$

7. $3\frac{5}{14} - \frac{11}{14} = \frac{47}{14} - \frac{11}{14} = 2\frac{4}{7}$

8. $3\frac{3}{11} - 1\frac{9}{11} = \frac{36}{11} - \frac{20}{11} = 1\frac{5}{11}$

9. $4\frac{5}{24} - 2\frac{17}{24} = \frac{101}{24} - \frac{65}{24} = 1\frac{1}{2}$

10. $6\frac{5}{14} - 4\frac{11}{14} = \frac{89}{14} - \frac{67}{14} = 1\frac{4}{7}$

11. $5\frac{3}{16} - 3\frac{15}{16} = \frac{83}{16} - \frac{63}{16} = 1\frac{1}{4}$

12. $5\frac{1}{12} - 1\frac{11}{12} = \frac{61}{12} - \frac{23}{12} = 3\frac{1}{6}$

13. $6\frac{7}{19} - 4\frac{18}{19} = \frac{121}{19} - \frac{94}{19} = 1\frac{8}{19}$

14. $4\frac{2}{13} - 2\frac{12}{13} = \frac{54}{13} - \frac{38}{13} = 1\frac{3}{13}$

15. $4\frac{3}{23} - \frac{11}{23} = \frac{95}{23} - \frac{11}{23} = 3\frac{15}{23}$

16. $4\frac{3}{28} - 1\frac{19}{28} = \frac{115}{28} - \frac{47}{28} = 2\frac{3}{7}$

17. $3\frac{9}{25} - 1\frac{3}{25} = \frac{84}{25} - \frac{28}{25} = 2\frac{6}{25}$

18. $5\frac{4}{19} - 1\frac{16}{19} = \frac{99}{19} - \frac{35}{19} = 3\frac{7}{19}$

Lesson 6-8 Subtracting fractions with the same denominators

Subtract fractions and write the answers in simplest form.

1. $2\frac{5}{8} - \frac{7}{8} = \frac{21}{8} - \frac{7}{8} = 1\frac{3}{4}$

2. $2\frac{1}{7} - 1\frac{6}{7} = \frac{15}{7} - \frac{13}{7} = \frac{2}{7}$

3. $3\frac{1}{5} - 2\frac{4}{5} = \frac{16}{5} - \frac{14}{5} = \frac{2}{5}$

4. $4\frac{2}{9} - \frac{7}{9} = \frac{38}{9} - \frac{7}{9} = 3\frac{4}{9}$

5. $4\frac{3}{13} - 1\frac{9}{13} = \frac{55}{13} - \frac{22}{13} = 2\frac{7}{13}$

6. $3\frac{5}{14} - 1\frac{11}{14} = \frac{47}{14} - \frac{25}{14} = 1\frac{4}{7}$

7. $3\frac{4}{25} - 1\frac{17}{25} = \frac{79}{25} - \frac{42}{25} = 1\frac{12}{25}$

8. $6\frac{1}{11} - 4\frac{9}{11} = \frac{67}{11} - \frac{53}{11} = 1\frac{3}{11}$

9. $6\frac{3}{17} - 3\frac{9}{17} = \frac{105}{17} - \frac{60}{17} = 2\frac{11}{17}$

10. $5\frac{2}{12} - 4\frac{10}{12} = \frac{62}{12} - \frac{58}{12} = \frac{1}{3}$

11. $5\frac{3}{20} - 1\frac{14}{20} = \frac{103}{20} - \frac{34}{20} = 3\frac{9}{20}$

12. $6\frac{3}{19} - 2\frac{15}{19} = \frac{117}{19} - \frac{53}{19} = 3\frac{7}{19}$

13. $6\frac{2}{17} - 5\frac{5}{17} = \frac{104}{17} - \frac{90}{17} = \frac{14}{17}$

14. $5\frac{3}{16} - 1\frac{7}{16} = \frac{83}{16} - \frac{23}{16} = 3\frac{3}{4}$

15. $4\frac{6}{11} - 2\frac{10}{11} = \frac{50}{11} - \frac{32}{11} = 1\frac{7}{11}$

16. $4\frac{5}{23} - 2\frac{19}{23} = \frac{97}{23} - \frac{65}{23} = 1\frac{9}{23}$

17. $5\frac{2}{12} - 2\frac{10}{12} = \frac{62}{12} - \frac{34}{12} = 2\frac{1}{3}$

18. $6\frac{2}{15} - 3\frac{13}{15} = \frac{92}{15} - \frac{58}{15} = 2\frac{4}{15}$

Lesson 6-9 Subtracting fractions with the same denominators

Subtract fractions and write the answers in simplest form.

1. $3\frac{1}{4} - \frac{5}{4} = \frac{13}{4} - \frac{5}{4} = 2$

2. $4\frac{1}{9} - 3\frac{7}{9} = \frac{37}{9} - \frac{34}{9} = \frac{1}{3}$

3. $4\frac{1}{8} - 2\frac{9}{8} = \frac{33}{8} - \frac{25}{8} = 1$

4. $7\frac{3}{5} - 4\frac{4}{5} = \frac{38}{5} - \frac{24}{5} = 2\frac{4}{5}$

5. $2\frac{3}{17} - 1\frac{19}{17} = \frac{37}{17} - \frac{36}{17} = \frac{1}{17}$

6. $5\frac{3}{20} - 1\frac{13}{20} = \frac{103}{20} - \frac{33}{20} = 3\frac{1}{2}$

7. $4\frac{5}{14} - \frac{9}{14} = \frac{61}{14} - \frac{9}{14} = 3\frac{5}{7}$

8. $4\frac{5}{17} - 2\frac{14}{17} = \frac{73}{17} - \frac{48}{17} = 1\frac{8}{17}$

9. $5\frac{9}{25} - 1\frac{17}{25} = \frac{134}{25} - \frac{42}{25} = 3\frac{17}{25}$

10. $6\frac{2}{19} - 3\frac{20}{19} = \frac{116}{19} - \frac{77}{19} = 2\frac{1}{19}$

11. $6\frac{7}{21} - \frac{17}{21} = \frac{133}{21} - \frac{17}{21} = 5\frac{11}{21}$

12. $5\frac{6}{13} - 2\frac{8}{13} = \frac{71}{13} - \frac{34}{13} = 2\frac{11}{13}$

13. $7\frac{5}{14} - 3\frac{13}{14} = \frac{103}{14} - \frac{55}{14} = 3\frac{3}{7}$

14. $6\frac{4}{25} - 4\frac{21}{25} = \frac{154}{25} - \frac{121}{25} = 1\frac{8}{25}$

15. $6\frac{4}{30} - 4\frac{21}{30} = \frac{184}{30} - \frac{141}{30} = 1\frac{13}{30}$

16. $5\frac{6}{15} - 2\frac{13}{15} = \frac{81}{15} - \frac{43}{15} = 2\frac{8}{15}$

17. $5\frac{3}{16} - 1\frac{19}{16} = \frac{83}{16} - \frac{35}{16} = 3$

18. $7\frac{5}{12} - 5\frac{11}{12} = \frac{89}{12} - \frac{71}{12} = 1\frac{1}{2}$

Lesson 6-10 Subtracting fractions with the same denominators

Subtract fractions and write the answers in simplest form.

1. $5\frac{3}{5} - 1\frac{8}{5} = \frac{28}{5} - \frac{13}{5} = 3$

2. $7\frac{1}{9} - 3\frac{8}{9} = \frac{64}{9} - \frac{35}{9} = 3\frac{2}{9}$

3. $6\frac{5}{6} - \frac{9}{6} = \frac{41}{6} - \frac{9}{6} = 5\frac{1}{3}$

4. $6\frac{3}{7} - 3\frac{10}{7} = \frac{45}{7} - \frac{31}{7} = 2$

5. $3\frac{3}{26} - 1\frac{15}{26} = \frac{81}{26} - \frac{41}{26} = 1\frac{7}{13}$

6. $4\frac{2}{11} - 1\frac{8}{11} = \frac{46}{11} - \frac{19}{11} = 2\frac{5}{11}$

7. $4\frac{2}{17} - 2\frac{14}{17} = \frac{70}{17} - \frac{48}{17} = 1\frac{5}{17}$

8. $5\frac{4}{15} - 2\frac{11}{15} = \frac{79}{15} - \frac{41}{15} = 2\frac{8}{15}$

9. $6\frac{4}{21} - 2\frac{22}{21} = \frac{130}{21} - \frac{64}{21} = 3\frac{1}{7}$

10. $5\frac{3}{16} - 4\frac{13}{16} = \frac{83}{16} - \frac{77}{16} = \frac{3}{8}$

11. $5\frac{5}{12} - 1\frac{17}{12} = \frac{65}{12} - \frac{29}{12} = 3$

12. $4\frac{5}{17} - 1\frac{14}{17} = \frac{73}{17} - \frac{31}{17} = 2\frac{8}{17}$

13. $4\frac{2}{19} - 3\frac{9}{19} = \frac{78}{19} - \frac{66}{19} = \frac{12}{19}$

14. $6\frac{6}{25} - 2\frac{16}{25} = \frac{156}{25} - \frac{66}{25} = 3\frac{3}{5}$

15. $5\frac{3}{16} - 2\frac{18}{16} = \frac{83}{16} - \frac{50}{16} = 2\frac{1}{16}$

16. $5\frac{2}{13} - 4\frac{9}{13} = \frac{67}{13} - \frac{61}{13} = \frac{6}{13}$

17. $4\frac{4}{30} - 1\frac{21}{30} = \frac{124}{30} - \frac{51}{30} = 2\frac{13}{30}$

18. $6\frac{5}{12} - 2\frac{7}{12} = \frac{77}{12} - \frac{31}{12} = 3\frac{5}{6}$

Lesson 7-1 Adding fractions with different denominators

Add fractions and write the answers in simplest form.

1. $\frac{1}{2} + \frac{1}{4} = \frac{2}{4} + \frac{1}{4} = \frac{3}{4}$ 2. $\frac{2}{5} + \frac{3}{10} = \frac{4}{10} + \frac{3}{10} = \frac{7}{10}$

3. $\frac{1}{2} + \frac{1}{6} = \frac{3}{6} + \frac{1}{6} = \frac{2}{3}$ 4. $\frac{1}{6} + \frac{5}{12} = \frac{2}{12} + \frac{5}{12} = \frac{7}{12}$

5. $\frac{2}{3} + \frac{1}{6} = \frac{4}{6} + \frac{1}{6} = \frac{5}{6}$ 6. $\frac{3}{4} + \frac{7}{12} = \frac{9}{12} + \frac{7}{12} = 1\frac{1}{3}$

7. $\frac{3}{5} + \frac{1}{10} = \frac{6}{10} + \frac{1}{10} = \frac{7}{10}$ 8. $\frac{2}{7} + \frac{5}{14} = \frac{4}{14} + \frac{5}{14} = \frac{9}{14}$

9. $\frac{2}{7} + \frac{3}{14} = \frac{4}{14} + \frac{3}{14} = \frac{7}{14}$ 10. $\frac{2}{3} + \frac{7}{9} = \frac{6}{9} + \frac{7}{9} = 1\frac{4}{9}$

11. $\frac{1}{2} + \frac{3}{8} = \frac{4}{8} + \frac{3}{8} = \frac{7}{8}$ 12. $\frac{1}{4} + \frac{5}{8} = \frac{2}{8} + \frac{5}{8} = \frac{7}{8}$

13. $\frac{2}{3} + \frac{4}{9} = \frac{6}{9} + \frac{4}{9} = 1\frac{1}{9}$ 14. $\frac{1}{2} + \frac{7}{8} = \frac{4}{8} + \frac{7}{8} = 1\frac{3}{8}$

15. $\frac{1}{2} + \frac{7}{10} = \frac{5}{10} + \frac{7}{10} = 1\frac{1}{5}$ 16. $\frac{2}{5} + \frac{11}{15} = \frac{6}{15} + \frac{11}{15} = 1\frac{2}{15}$

17. $\frac{3}{4} + \frac{7}{8} = \frac{6}{8} + \frac{7}{8} = 1\frac{5}{8}$ 18. $\frac{3}{8} + \frac{15}{16} = \frac{6}{16} + \frac{15}{16} = 1\frac{5}{16}$

Lesson 7-2 Adding fractions with different denominators

Add fractions and write the answers in simplest form.

1. $\frac{3}{4} + \frac{5}{8} = \frac{6}{8} + \frac{5}{8} = 1\frac{3}{8}$ 2. $\frac{2}{3} + \frac{5}{9} = \frac{6}{9} + \frac{5}{9} = 1\frac{2}{9}$

3. $\frac{2}{5} + \frac{3}{10} = \frac{4}{10} + \frac{3}{10} = \frac{7}{10}$ 4. $\frac{1}{2} + \frac{7}{12} = \frac{6}{12} + \frac{7}{12} = 1\frac{1}{12}$

5. $\frac{1}{4} + \frac{7}{12} = \frac{3}{12} + \frac{7}{12} = \frac{5}{6}$ 6. $\frac{1}{5} + \frac{6}{15} = \frac{3}{15} + \frac{6}{15} = \frac{3}{5}$

7. $\frac{5}{6} + \frac{7}{18} = \frac{15}{18} + \frac{7}{18} = 1\frac{2}{9}$ 8. $\frac{5}{12} + \frac{7}{24} = \frac{10}{24} + \frac{7}{24} = \frac{17}{24}$

9. $\frac{2}{7} + \frac{3}{14} = \frac{4}{14} + \frac{3}{14} = \frac{1}{2}$ 10. $\frac{1}{7} + \frac{11}{21} = \frac{3}{21} + \frac{11}{21} = \frac{2}{3}$

11. $\frac{4}{5} + \frac{11}{15} = \frac{12}{15} + \frac{11}{15} = 1\frac{8}{15}$ 12. $\frac{2}{5} + \frac{9}{10} = \frac{4}{10} + \frac{9}{10} = 1\frac{3}{10}$

13. $\frac{5}{9} + \frac{13}{18} = \frac{10}{18} + \frac{13}{18} = 1\frac{5}{18}$ 14. $\frac{1}{6} + \frac{13}{18} = \frac{3}{18} + \frac{13}{18} = \frac{8}{9}$

15. $\frac{3}{10} + \frac{9}{20} = \frac{6}{20} + \frac{9}{20} = \frac{3}{4}$ 16. $\frac{4}{5} + \frac{3}{20} = \frac{16}{20} + \frac{3}{20} = \frac{19}{20}$

17. $\frac{2}{5} + \frac{7}{20} = \frac{8}{20} + \frac{7}{20} = \frac{3}{4}$ 18. $\frac{1}{4} + \frac{15}{16} = \frac{4}{16} + \frac{15}{16} = 1\frac{3}{16}$

Lesson 7-3 Adding fractions with different denominators

Add fractions and write the answers in simplest form.

1. $\frac{1}{2} + \frac{1}{3} = \frac{3}{6} + \frac{2}{6} = \frac{5}{6}$ 2. $\frac{1}{3} + \frac{2}{5} = \frac{5}{15} + \frac{6}{15} = \frac{11}{15}$

3. $\frac{2}{3} + \frac{1}{4} = \frac{8}{12} + \frac{3}{12} = \frac{11}{12}$ 4. $\frac{2}{3} + \frac{3}{4} = \frac{8}{12} + \frac{9}{12} = 1\frac{5}{12}$

5. $\frac{1}{3} + \frac{1}{4} = \frac{4}{12} + \frac{3}{12} = \frac{7}{12}$ 6. $\frac{1}{2} + \frac{4}{5} = \frac{5}{10} + \frac{8}{10} = 1\frac{3}{10}$

7. $\frac{2}{3} + \frac{1}{6} = \frac{4}{6} + \frac{1}{6} = \frac{5}{6}$ 8. $\frac{1}{3} + \frac{7}{9} = \frac{3}{9} + \frac{7}{9} = 1\frac{1}{9}$

9. $\frac{2}{3} + \frac{2}{5} = \frac{10}{15} + \frac{6}{15} = 1\frac{1}{15}$ 10. $\frac{1}{4} + \frac{2}{5} = \frac{5}{20} + \frac{8}{20} = \frac{13}{20}$

11. $\frac{3}{10} + \frac{4}{20} = \frac{6}{20} + \frac{4}{20} = \frac{1}{2}$ 12. $\frac{3}{4} + \frac{7}{16} = \frac{12}{16} + \frac{7}{16} = 1\frac{3}{16}$

13. $\frac{3}{4} + \frac{2}{5} = \frac{15}{20} + \frac{8}{20} = 1\frac{3}{20}$ 14. $\frac{2}{7} + \frac{5}{14} = \frac{4}{14} + \frac{5}{14} = \frac{9}{14}$

15. $\frac{1}{3} + \frac{7}{15} = \frac{5}{15} + \frac{7}{15} = \frac{4}{5}$ 16. $\frac{1}{4} + \frac{5}{6} = \frac{3}{12} + \frac{10}{12} = 1\frac{1}{12}$

17. $\frac{2}{5} + \frac{7}{15} = \frac{6}{15} + \frac{7}{15} = \frac{13}{15}$ 18. $\frac{2}{3} + \frac{11}{12} = \frac{8}{12} + \frac{11}{12} = 1\frac{7}{12}$

Lesson 7-4 Adding fractions with different denominators

Add fractions and write the answers in simplest form.

1. $\frac{1}{2} + \frac{1}{5} = \frac{5}{10} + \frac{2}{10} = \frac{7}{10}$ 2. $\frac{1}{2} + \frac{3}{7} = \frac{7}{14} + \frac{6}{14} = \frac{13}{14}$

3. $\frac{2}{3} + \frac{5}{9} = \frac{6}{9} + \frac{5}{9} = 1\frac{2}{9}$ 4. $\frac{2}{3} + \frac{2}{7} = \frac{14}{21} + \frac{6}{21} = \frac{20}{21}$

5. $\frac{1}{2} + \frac{2}{7} = \frac{7}{14} + \frac{4}{14} = \frac{11}{14}$ 6. $\frac{1}{5} + \frac{11}{20} = \frac{4}{20} + \frac{11}{20} = \frac{3}{4}$

7. $\frac{2}{3} + \frac{3}{4} = \frac{8}{12} + \frac{9}{12} = 1\frac{5}{12}$ 8. $\frac{5}{9} + \frac{7}{18} = \frac{10}{18} + \frac{7}{18} = \frac{17}{18}$

9. $\frac{1}{2} + \frac{9}{14} = \frac{7}{14} + \frac{9}{14} = 1\frac{1}{7}$ 10. $\frac{5}{6} + \frac{2}{5} = \frac{25}{30} + \frac{12}{30} = 1\frac{7}{30}$

11. $\frac{1}{3} + \frac{2}{5} = \frac{5}{15} + \frac{6}{15} = \frac{11}{15}$ 12. $\frac{2}{3} + \frac{3}{4} = \frac{8}{12} + \frac{9}{12} = 1\frac{5}{12}$

13. $\frac{3}{4} + \frac{1}{6} = \frac{9}{12} + \frac{2}{12} = \frac{11}{12}$ 14. $\frac{3}{11} + \frac{5}{22} = \frac{6}{22} + \frac{5}{22} = \frac{1}{2}$

15. $\frac{1}{3} + \frac{5}{6} = \frac{2}{6} + \frac{5}{6} = 1\frac{1}{6}$ 16. $\frac{1}{4} + \frac{6}{7} = \frac{7}{28} + \frac{24}{28} = 1\frac{3}{28}$

17. $\frac{1}{4} + \frac{2}{5} = \frac{5}{20} + \frac{8}{20} = \frac{13}{20}$ 18. $\frac{2}{3} + \frac{7}{8} = \frac{16}{24} + \frac{21}{24} = 1\frac{13}{24}$

Lesson 7-5   Adding fractions with different denominators

Add fractions and write the answers in simplest form.

1. $\frac{1}{2}+\frac{1}{3}=\frac{3}{6}+\frac{2}{6}=\frac{5}{6}$

2. $\frac{2}{7}+\frac{5}{21}=\frac{6}{21}+\frac{5}{21}=\frac{11}{21}$

3. $\frac{2}{3}+\frac{1}{4}=\frac{8}{12}+\frac{3}{12}=\frac{11}{12}$

4. $\frac{1}{8}+\frac{2}{5}=\frac{5}{40}+\frac{16}{40}=\frac{21}{40}$

5. $\frac{1}{2}+\frac{4}{9}=\frac{9}{18}+\frac{8}{18}=\frac{17}{18}$

6. $\frac{7}{12}+\frac{7}{36}=\frac{21}{36}+\frac{7}{36}=\frac{7}{9}$

7. $\frac{1}{3}+\frac{5}{7}=\frac{7}{21}+\frac{15}{21}=1\frac{1}{21}$

8. $\frac{3}{8}+\frac{5}{6}=\frac{9}{24}+\frac{20}{24}=1\frac{5}{24}$

9. $\frac{2}{3}+\frac{1}{15}=\frac{10}{15}+\frac{1}{15}=\frac{11}{15}$

10. $\frac{2}{9}+\frac{2}{5}=\frac{10}{45}+\frac{18}{45}=\frac{28}{45}$

11. $\frac{3}{11}+\frac{7}{22}=\frac{6}{22}+\frac{7}{22}=\frac{13}{22}$

12. $\frac{1}{6}+\frac{13}{24}=\frac{4}{24}+\frac{13}{24}=\frac{17}{24}$

13. $\frac{2}{7}+\frac{1}{6}=\frac{12}{42}+\frac{7}{42}=\frac{19}{42}$

14. $\frac{5}{7}+\frac{3}{5}=\frac{25}{35}+\frac{21}{35}=1\frac{11}{35}$

15. $\frac{3}{4}+\frac{2}{5}=\frac{15}{20}+\frac{8}{20}=1\frac{3}{20}$

16. $\frac{5}{12}+\frac{5}{18}=\frac{15}{36}+\frac{10}{36}=\frac{25}{36}$

17. $\frac{4}{15}+\frac{11}{30}=\frac{8}{30}+\frac{11}{30}=\frac{19}{30}$

18. $\frac{2}{13}+\frac{9}{39}=\frac{6}{39}+\frac{9}{39}=\frac{5}{13}$

Lesson 7-6   Adding fractions with different denominators

Add fractions and write the answers in simplest form.

1. $\frac{1}{3}+\frac{3}{10}=\frac{10}{30}+\frac{9}{30}=\frac{19}{30}$

2. $\frac{6}{13}+\frac{7}{26}=\frac{12}{26}+\frac{7}{26}=\frac{19}{26}$

3. $\frac{1}{2}+\frac{4}{9}=\frac{9}{18}+\frac{8}{18}=\frac{17}{18}$

4. $\frac{4}{5}+\frac{3}{7}=\frac{28}{35}+\frac{15}{35}=1\frac{8}{35}$

5. $\frac{3}{4}+\frac{3}{5}=\frac{15}{20}+\frac{12}{20}=1\frac{7}{20}$

6. $\frac{3}{10}+\frac{17}{40}=\frac{12}{40}+\frac{17}{40}=\frac{29}{40}$

7. $\frac{2}{5}+\frac{9}{25}=\frac{10}{25}+\frac{9}{25}=\frac{19}{25}$

8. $\frac{5}{6}+\frac{3}{7}=\frac{35}{42}+\frac{18}{42}=1\frac{11}{42}$

9. $\frac{7}{11}+\frac{7}{33}=\frac{21}{33}+\frac{7}{33}=\frac{28}{33}$

10. $\frac{2}{5}+\frac{4}{9}=\frac{18}{45}+\frac{20}{45}=\frac{38}{45}$

11. $\frac{2}{7}+\frac{11}{28}=\frac{8}{28}+\frac{11}{28}=\frac{19}{28}$

12. $\frac{5}{11}+\frac{3}{44}=\frac{20}{44}+\frac{3}{44}=\frac{23}{44}$

13. $\frac{5}{9}+\frac{14}{27}=\frac{15}{27}+\frac{14}{27}=1\frac{2}{27}$

14. $\frac{7}{12}+\frac{4}{15}=\frac{35}{60}+\frac{16}{60}=\frac{51}{60}$

15. $\frac{5}{12}+\frac{13}{18}=\frac{15}{36}+\frac{26}{36}=1\frac{5}{36}$

16. $\frac{3}{10}+\frac{5}{12}=\frac{18}{60}+\frac{25}{60}=\frac{43}{60}$

17. $\frac{3}{5}+\frac{4}{9}=\frac{27}{45}+\frac{20}{45}=1\frac{2}{45}$

18. $\frac{1}{6}+\frac{23}{36}=\frac{6}{36}+\frac{23}{36}=\frac{29}{36}$

Lesson 7-7   Adding fractions with different denominators

Add fractions and write the answers in simplest form.

1. $\frac{1}{3}+\frac{3}{5}=\frac{5}{15}+\frac{9}{15}=\frac{14}{15}$

2. $\frac{7}{12}+\frac{13}{48}=\frac{28}{48}+\frac{13}{48}=\frac{41}{48}$

3. $\frac{3}{11}+\frac{10}{33}=\frac{9}{33}+\frac{10}{33}=\frac{19}{33}$

4. $\frac{2}{9}+\frac{23}{45}=\frac{10}{45}+\frac{23}{45}=\frac{11}{15}$

5. $\frac{3}{4}+\frac{2}{7}=\frac{21}{28}+\frac{8}{28}=1\frac{1}{28}$

6. $\frac{7}{16}+\frac{13}{24}=\frac{21}{48}+\frac{26}{48}=\frac{47}{48}$

7. $\frac{5}{6}+\frac{4}{9}=\frac{45}{54}+\frac{24}{54}=1\frac{5}{18}$

8. $\frac{7}{15}+\frac{26}{45}=\frac{21}{45}+\frac{26}{45}=1\frac{2}{45}$

9. $\frac{3}{5}+\frac{3}{4}=\frac{12}{20}+\frac{15}{20}=1\frac{7}{20}$

10. $\frac{4}{9}+\frac{7}{8}=\frac{32}{72}+\frac{63}{72}=1\frac{23}{72}$

11. $\frac{3}{14}+\frac{9}{42}=\frac{9}{42}+\frac{9}{42}=\frac{3}{7}$

12. $\frac{3}{5}+\frac{5}{6}=\frac{18}{30}+\frac{25}{30}=1\frac{13}{30}$

13. $\frac{7}{12}+\frac{23}{60}=\frac{35}{60}+\frac{23}{60}=\frac{29}{30}$

14. $\frac{7}{12}+\frac{5}{18}=\frac{21}{36}+\frac{10}{36}=\frac{31}{36}$

15. $\frac{3}{4}+\frac{8}{9}=\frac{27}{36}+\frac{32}{36}=1\frac{23}{36}$

16. $\frac{9}{14}+\frac{2}{35}=\frac{45}{70}+\frac{4}{70}=\frac{7}{10}$

17. $\frac{6}{7}+\frac{7}{8}=\frac{48}{56}+\frac{49}{56}=1\frac{41}{56}$

18. $\frac{5}{18}+\frac{22}{45}=\frac{25}{90}+\frac{44}{90}=\frac{23}{30}$

Lesson 7-8   Adding fractions with different denominators

Add fractions and write the answers in simplest form.

1. $\frac{1}{2}+\frac{3}{7}=\frac{7}{14}+\frac{6}{14}=\frac{13}{14}$

2. $\frac{4}{5}+\frac{3}{4}=\frac{16}{20}+\frac{15}{20}=1\frac{11}{20}$

3. $\frac{2}{3}+\frac{3}{8}=\frac{16}{24}+\frac{9}{24}=1\frac{1}{24}$

4. $\frac{2}{7}+\frac{5}{8}=\frac{16}{56}+\frac{35}{56}=\frac{51}{56}$

5. $\frac{1}{2}+\frac{7}{11}=\frac{11}{22}+\frac{14}{22}=1\frac{3}{22}$

6. $\frac{3}{5}+\frac{5}{7}=\frac{21}{35}+\frac{25}{35}=1\frac{11}{35}$

7. $\frac{1}{3}+\frac{10}{21}=\frac{7}{21}+\frac{10}{21}=\frac{17}{21}$

8. $\frac{1}{4}+\frac{2}{3}=\frac{3}{12}+\frac{8}{12}=\frac{11}{12}$

9. $\frac{7}{12}+\frac{11}{48}=\frac{28}{48}+\frac{11}{48}=\frac{39}{48}$

10. $\frac{9}{14}+\frac{11}{21}=\frac{27}{42}+\frac{22}{42}=1\frac{1}{6}$

11. $\frac{5}{12}+\frac{7}{9}=\frac{15}{36}+\frac{28}{36}=1\frac{7}{36}$

12. $\frac{7}{15}+\frac{22}{45}=\frac{21}{45}+\frac{22}{45}=\frac{43}{45}$

13. $\frac{3}{8}+\frac{11}{12}=\frac{9}{24}+\frac{22}{24}=1\frac{7}{24}$

14. $\frac{5}{16}+\frac{5}{24}=\frac{15}{48}+\frac{10}{48}=\frac{25}{48}$

15. $\frac{4}{5}+\frac{6}{7}=\frac{28}{35}+\frac{30}{35}=1\frac{23}{35}$

16. $\frac{7}{12}+\frac{13}{18}=\frac{21}{36}+\frac{26}{36}=1\frac{11}{36}$

17. $\frac{5}{6}+\frac{8}{9}=\frac{45}{54}+\frac{48}{54}=1\frac{13}{18}$

18. $\frac{13}{18}+\frac{4}{45}=\frac{65}{90}+\frac{8}{90}=\frac{73}{90}$

Lesson 7-9 Adding fractions with different denominators

Add fractions and write the answers in simplest form.

1. $1\frac{1}{2} + \frac{3}{7} = \frac{21}{14} + \frac{6}{14} = 1\frac{13}{14}$
2. $\frac{7}{12} + 1\frac{5}{9} = \frac{21}{36} + \frac{56}{36} = 2\frac{5}{36}$
3. $1\frac{1}{3} + \frac{4}{5} = \frac{20}{15} + \frac{12}{15} = 2\frac{2}{15}$
4. $1\frac{1}{18} + \frac{7}{12} = \frac{38}{36} + \frac{21}{36} = 1\frac{23}{36}$
5. $\frac{2}{3} + 1\frac{3}{14} = \frac{28}{42} + \frac{51}{42} = 1\frac{37}{42}$
6. $1\frac{3}{14} + \frac{22}{35} = \frac{85}{70} + \frac{44}{70} = 1\frac{59}{70}$
7. $\frac{3}{4} + 1\frac{2}{9} = \frac{27}{36} + \frac{44}{36} = 1\frac{35}{36}$
8. $\frac{5}{16} + 1\frac{13}{40} = \frac{25}{80} + \frac{106}{80} = 1\frac{51}{80}$
9. $\frac{4}{5} + \frac{5}{9} = \frac{36}{45} + \frac{25}{45} = 1\frac{16}{45}$
10. $1\frac{5}{6} + 1\frac{3}{7} = \frac{77}{42} + \frac{60}{42} = 3\frac{11}{42}$
11. $1\frac{1}{12} + 1\frac{1}{36} = \frac{39}{36} + \frac{37}{36} = 2\frac{1}{9}$
12. $1\frac{6}{15} + 1\frac{23}{60} = \frac{84}{60} + \frac{83}{60} = 2\frac{47}{60}$
13. $1\frac{3}{13} + 1\frac{7}{52} = \frac{64}{52} + \frac{59}{52} = 2\frac{19}{52}$
14. $\frac{3}{10} + 1\frac{9}{14} = \frac{21}{70} + \frac{115}{70} = 1\frac{33}{35}$
15. $\frac{7}{8} + 1\frac{11}{12} = \frac{21}{24} + \frac{46}{24} = 2\frac{19}{24}$
16. $1\frac{3}{18} + \frac{17}{24} = \frac{84}{72} + \frac{51}{72} = 1\frac{7}{8}$
17. $1\frac{5}{12} + \frac{9}{32} = \frac{136}{96} + \frac{27}{96} = 1\frac{67}{96}$
18. $1\frac{11}{18} + 1\frac{5}{27} = \frac{87}{54} + \frac{64}{54} = 2\frac{43}{54}$

Lesson 7-10 Adding fractions with different denominators

Add fractions and write the answers in simplest form.

1. $1\frac{1}{2} + 1\frac{3}{11} = \frac{33}{22} + \frac{28}{22} = 2\frac{17}{22}$
2. $\frac{4}{5} + 1\frac{2}{9} = \frac{36}{45} + \frac{55}{45} = 2\frac{1}{45}$
3. $\frac{3}{4} + 1\frac{4}{9} = \frac{27}{36} + \frac{52}{36} = 1\frac{7}{36}$
4. $1\frac{6}{7} + \frac{4}{9} = \frac{117}{63} + \frac{28}{63} = 2\frac{19}{63}$
5. $1\frac{1}{3} + \frac{5}{7} = \frac{28}{21} + \frac{15}{21} = 2\frac{1}{21}$
6. $1\frac{2}{3} + 1\frac{15}{16} = \frac{80}{48} + \frac{93}{48} = 3\frac{29}{48}$
7. $1\frac{3}{5} + 1\frac{3}{8} = \frac{64}{40} + \frac{55}{40} = 2\frac{39}{40}$
8. $1\frac{3}{4} + \frac{16}{21} = \frac{147}{84} + \frac{64}{84} = 2\frac{43}{84}$
9. $1\frac{1}{6} + \frac{8}{9} = \frac{63}{54} + \frac{48}{54} = 2\frac{1}{18}$
10. $\frac{9}{14} + 1\frac{22}{56} = \frac{36}{56} + \frac{78}{56} = 2\frac{1}{28}$
11. $\frac{5}{12} + 1\frac{23}{60} = \frac{25}{60} + \frac{83}{60} = 1\frac{4}{5}$
12. $1\frac{13}{26} + 1\frac{24}{39} = \frac{243}{78} = 3\frac{3}{26}$
13. $1\frac{7}{22} + 1\frac{12}{33} = \frac{87}{66} + \frac{90}{66} = 2\frac{15}{22}$
14. $1\frac{7}{8} + 1\frac{11}{12} = \frac{45}{24} + \frac{46}{24} = 3\frac{19}{24}$
15. $1\frac{7}{13} + 1\frac{10}{65} = \frac{100}{65} + \frac{75}{65} = 2\frac{9}{13}$
16. $1\frac{7}{8} + \frac{9}{14} = \frac{105}{56} + \frac{36}{56} = 2\frac{29}{56}$
17. $\frac{5}{8} + 1\frac{9}{14} = \frac{35}{56} + \frac{92}{56} = 2\frac{15}{56}$
18. $\frac{11}{18} + 1\frac{13}{27} = \frac{33}{54} + \frac{80}{54} = 2\frac{5}{54}$

Lesson 8-1 Subtracting fractions with different denominators

Subtract fractions and write the answers in simplest form.

1. $\frac{1}{2} - \frac{1}{3} = \frac{3}{6} - \frac{2}{6} = \frac{1}{6}$
2. $\frac{2}{3} - \frac{1}{4} = \frac{8}{12} - \frac{3}{12} = \frac{5}{12}$
3. $\frac{1}{2} - \frac{2}{5} = \frac{5}{10} - \frac{4}{10} = \frac{1}{10}$
4. $\frac{1}{3} - \frac{1}{5} = \frac{5}{15} - \frac{3}{15} = \frac{2}{15}$
5. $\frac{1}{2} - \frac{3}{10} = \frac{5}{10} - \frac{3}{10} = \frac{1}{5}$
6. $\frac{2}{3} - \frac{1}{6} = \frac{4}{6} - \frac{1}{6} = \frac{1}{2}$
7. $\frac{3}{4} - \frac{3}{8} = \frac{12}{16} - \frac{6}{16} = \frac{3}{8}$
8. $\frac{2}{3} - \frac{1}{7} = \frac{14}{21} - \frac{3}{21} = \frac{11}{21}$
9. $\frac{1}{4} - \frac{1}{12} = \frac{3}{12} - \frac{1}{12} = \frac{1}{6}$
10. $\frac{2}{5} - \frac{1}{10} = \frac{4}{10} - \frac{1}{10} = \frac{3}{10}$
11. $\frac{5}{6} - \frac{7}{12} = \frac{10}{12} - \frac{7}{12} = \frac{1}{4}$
12. $\frac{4}{5} - \frac{7}{15} = \frac{12}{15} - \frac{7}{15} = \frac{1}{3}$
13. $\frac{5}{6} - \frac{5}{8} = \frac{20}{24} - \frac{15}{24} = \frac{5}{24}$
14. $\frac{3}{5} - \frac{1}{6} = \frac{18}{30} - \frac{5}{30} = \frac{13}{30}$
15. $\frac{6}{7} - \frac{9}{14} = \frac{12}{14} - \frac{9}{14} = \frac{3}{14}$
16. $\frac{1}{5} - \frac{1}{8} = \frac{8}{40} - \frac{5}{40} = \frac{3}{40}$
17. $\frac{3}{7} - \frac{8}{21} = \frac{9}{21} - \frac{8}{21} = \frac{1}{21}$
18. $\frac{7}{8} - \frac{7}{12} = \frac{21}{24} - \frac{14}{24} = \frac{7}{24}$

Lesson 8-2 Subtracting fractions with different denominators

Subtract fractions and write the answers in simplest form.

1. $\frac{1}{2} - \frac{1}{5} = \frac{5}{10} - \frac{2}{10} = \frac{3}{10}$
2. $\frac{2}{3} - \frac{3}{7} = \frac{14}{21} - \frac{9}{21} = \frac{5}{21}$
3. $\frac{2}{5} - \frac{1}{6} = \frac{12}{30} - \frac{5}{30} = \frac{7}{30}$
4. $\frac{3}{4} - \frac{1}{6} = \frac{9}{12} - \frac{2}{12} = \frac{7}{12}$
5. $\frac{2}{3} - \frac{1}{4} = \frac{8}{12} - \frac{3}{12} = \frac{5}{12}$
6. $\frac{7}{9} - \frac{5}{18} = \frac{14}{18} - \frac{5}{18} = \frac{1}{2}$
7. $\frac{5}{8} - \frac{5}{12} = \frac{15}{24} - \frac{10}{24} = \frac{5}{24}$
8. $\frac{7}{10} - \frac{11}{30} = \frac{21}{30} - \frac{11}{30} = \frac{1}{3}$
9. $\frac{3}{4} - \frac{3}{7} = \frac{21}{28} - \frac{12}{28} = \frac{9}{28}$
10. $\frac{3}{4} - \frac{5}{16} = \frac{12}{16} - \frac{5}{16} = \frac{7}{16}$
11. $\frac{3}{4} - \frac{1}{6} = \frac{9}{12} - \frac{2}{12} = \frac{7}{12}$
12. $\frac{4}{5} - \frac{5}{7} = \frac{28}{35} - \frac{25}{35} = \frac{3}{35}$
13. $\frac{4}{5} - \frac{9}{20} = \frac{16}{20} - \frac{9}{20} = \frac{7}{20}$
14. $\frac{5}{6} - \frac{3}{7} = \frac{35}{42} - \frac{18}{42} = \frac{17}{42}$
15. $\frac{5}{6} - \frac{11}{18} = \frac{15}{18} - \frac{11}{18} = \frac{2}{9}$
16. $\frac{3}{7} - \frac{2}{9} = \frac{27}{63} - \frac{14}{63} = \frac{13}{63}$
17. $\frac{7}{9} - \frac{5}{12} = \frac{28}{36} - \frac{15}{36} = \frac{13}{36}$
18. $\frac{7}{8} - \frac{2}{9} = \frac{63}{72} - \frac{16}{72} = \frac{47}{72}$

Lesson 8-3 Subtracting fractions with different denominators

Subtract fractions and write the answers in simplest form.

1. $\frac{1}{2} - \frac{2}{7} = \frac{7}{14} - \frac{4}{14} = \frac{3}{14}$
2. $\frac{3}{4} - \frac{3}{10} = \frac{15}{20} - \frac{6}{20} = \frac{9}{20}$

3. $\frac{3}{4} - \frac{2}{5} = \frac{15}{20} - \frac{8}{20} = \frac{7}{20}$
4. $\frac{4}{5} - \frac{2}{9} = \frac{36}{45} - \frac{10}{45} = \frac{26}{45}$

5. $\frac{2}{3} - \frac{3}{7} = \frac{14}{21} - \frac{9}{21} = \frac{5}{21}$
6. $\frac{11}{13} - \frac{9}{39} = \frac{33}{39} - \frac{9}{39} = \frac{8}{13}$

7. $\frac{5}{6} - \frac{2}{9} = \frac{15}{18} - \frac{4}{18} = \frac{11}{18}$
8. $\frac{11}{12} - \frac{7}{16} = \frac{44}{48} - \frac{21}{48} = \frac{23}{48}$

9. $\frac{1}{8} - \frac{1}{24} = \frac{3}{24} - \frac{1}{24} = \frac{1}{12}$
10. $\frac{9}{14} - \frac{8}{35} = \frac{45}{70} - \frac{16}{70} = \frac{29}{70}$

11. $\frac{3}{4} - \frac{2}{9} = \frac{27}{36} - \frac{8}{36} = \frac{19}{36}$
12. $\frac{4}{15} - \frac{1}{30} = \frac{8}{30} - \frac{1}{30} = \frac{7}{30}$

13. $\frac{3}{11} - \frac{1}{33} = \frac{9}{33} - \frac{1}{33} = \frac{8}{33}$
14. $\frac{15}{16} - \frac{9}{32} = \frac{30}{32} - \frac{9}{32} = \frac{21}{32}$

15. $\frac{7}{12} - \frac{5}{18} = \frac{21}{36} - \frac{10}{36} = \frac{11}{36}$
16. $\frac{9}{14} - \frac{8}{21} = \frac{27}{42} - \frac{16}{42} = \frac{11}{42}$

17. $\frac{7}{9} - \frac{1}{12} = \frac{28}{36} - \frac{3}{36} = \frac{25}{36}$
18. $\frac{15}{16} - \frac{13}{24} = \frac{45}{48} - \frac{26}{48} = \frac{19}{48}$

Lesson 8-4 Subtracting fractions with different denominators

Subtract fractions and write the answers in simplest form.

1. $\frac{2}{3} - \frac{4}{9} = \frac{18}{27} - \frac{12}{27} = \frac{2}{9}$
2. $\frac{5}{6} - \frac{7}{24} = \frac{20}{24} - \frac{7}{24} = \frac{13}{24}$

3. $\frac{1}{6} - \frac{1}{9} = \frac{3}{18} - \frac{2}{18} = \frac{1}{18}$
4. $\frac{2}{3} - \frac{1}{8} = \frac{16}{24} - \frac{3}{24} = \frac{13}{24}$

5. $\frac{7}{8} - \frac{13}{32} = \frac{28}{32} - \frac{13}{32} = \frac{15}{32}$
6. $\frac{7}{9} - \frac{5}{12} = \frac{28}{36} - \frac{15}{36} = \frac{13}{36}$

7. $\frac{5}{6} - \frac{9}{14} = \frac{35}{42} - \frac{27}{42} = \frac{4}{21}$
8. $\frac{9}{10} - \frac{11}{25} = \frac{45}{50} - \frac{22}{50} = \frac{23}{50}$

9. $\frac{1}{2} - \frac{1}{7} = \frac{7}{14} - \frac{2}{14} = \frac{5}{14}$
10. $\frac{9}{11} - \frac{25}{44} = \frac{36}{44} - \frac{25}{44} = \frac{1}{4}$

11. $\frac{11}{12} - \frac{9}{16} = \frac{44}{48} - \frac{27}{48} = \frac{17}{48}$
12. $\frac{13}{18} - \frac{11}{24} = \frac{52}{72} - \frac{33}{72} = \frac{19}{72}$

13. $\frac{5}{8} - \frac{1}{9} = \frac{45}{72} - \frac{8}{72} = \frac{37}{72}$
14. $\frac{15}{16} - \frac{21}{40} = \frac{75}{80} - \frac{42}{80} = \frac{33}{80}$

15. $\frac{11}{12} - \frac{8}{9} = \frac{33}{36} - \frac{32}{36} = \frac{1}{36}$
16. $\frac{7}{12} - \frac{17}{60} = \frac{35}{60} - \frac{17}{60} = \frac{3}{10}$

17. $\frac{13}{18} - \frac{13}{27} = \frac{39}{54} - \frac{26}{54} = \frac{13}{54}$
18. $\frac{13}{24} - \frac{11}{32} = \frac{52}{96} - \frac{33}{96} = \frac{19}{96}$

Lesson 8-5 Subtracting fractions with different denominators

Subtract fractions and write the answers in simplest form.

1. $\frac{3}{5} - \frac{1}{6} = \frac{18}{30} - \frac{5}{30} = \frac{13}{30}$
2. $\frac{3}{4} - \frac{3}{7} = \frac{21}{28} - \frac{12}{28} = \frac{9}{28}$

3. $\frac{1}{2} - \frac{5}{11} = \frac{11}{22} - \frac{10}{22} = \frac{1}{22}$
4. $\frac{1}{4} - \frac{1}{14} = \frac{7}{28} - \frac{2}{28} = \frac{5}{28}$

5. $\frac{5}{6} - \frac{4}{9} = \frac{15}{18} - \frac{8}{18} = \frac{7}{18}$
6. $\frac{5}{6} - \frac{11}{18} = \frac{30}{36} - \frac{22}{36} = \frac{2}{9}$

7. $\frac{5}{6} - \frac{8}{21} = \frac{35}{42} - \frac{16}{42} = \frac{19}{42}$
8. $\frac{13}{15} - \frac{11}{18} = \frac{78}{90} - \frac{55}{90} = \frac{23}{90}$

9. $\frac{3}{4} - \frac{11}{16} = \frac{24}{32} - \frac{22}{32} = \frac{1}{16}$
10. $\frac{5}{8} - \frac{7}{24} = \frac{15}{24} - \frac{7}{24} = \frac{1}{3}$

11. $\frac{3}{8} - \frac{1}{20} = \frac{15}{40} - \frac{2}{40} = \frac{13}{40}$
12. $\frac{7}{8} - \frac{15}{28} = \frac{49}{56} - \frac{30}{56} = \frac{19}{56}$

13. $\frac{11}{12} - \frac{37}{72} = \frac{66}{72} - \frac{37}{72} = \frac{29}{72}$
14. $\frac{10}{13} - \frac{31}{52} = \frac{40}{52} - \frac{31}{52} = \frac{9}{52}$

15. $\frac{10}{13} - \frac{19}{78} = \frac{60}{78} - \frac{19}{78} = \frac{41}{78}$
16. $\frac{13}{16} - \frac{51}{96} = \frac{78}{96} - \frac{51}{96} = \frac{9}{32}$

17. $\frac{14}{15} - \frac{28}{45} = \frac{42}{45} - \frac{28}{45} = \frac{14}{45}$
18. $\frac{21}{26} - \frac{19}{39} = \frac{63}{78} - \frac{38}{78} = \frac{25}{78}$

Lesson 8-6 Subtracting fractions with different denominators

Subtract fractions and write the answers in simplest form.

1. $\frac{2}{3} - \frac{11}{21} = \frac{14}{21} - \frac{11}{21} = \frac{1}{7}$
2. $\frac{3}{4} - \frac{9}{20} = \frac{15}{20} - \frac{9}{20} = \frac{3}{10}$

3. $\frac{3}{4} - \frac{3}{5} = \frac{15}{20} - \frac{12}{20} = \frac{3}{20}$
4. $\frac{2}{3} - \frac{5}{8} = \frac{16}{24} - \frac{15}{24} = \frac{1}{24}$

5. $\frac{2}{3} - \frac{1}{6} = \frac{4}{6} - \frac{1}{6} = \frac{1}{2}$
6. $\frac{4}{5} - \frac{16}{25} = \frac{20}{25} - \frac{16}{25} = \frac{4}{25}$

7. $\frac{11}{12} - \frac{13}{36} = \frac{33}{36} - \frac{13}{36} = \frac{5}{9}$
8. $\frac{5}{6} - \frac{7}{9} = \frac{15}{18} - \frac{14}{18} = \frac{1}{18}$

9. $\frac{11}{15} - \frac{17}{45} = \frac{33}{45} - \frac{17}{45} = \frac{16}{45}$
10. $\frac{9}{11} - \frac{31}{44} = \frac{36}{44} - \frac{31}{44} = \frac{5}{44}$

11. $\frac{7}{24} - \frac{1}{72} = \frac{21}{72} - \frac{1}{72} = \frac{5}{18}$
12. $\frac{12}{14} - \frac{13}{35} = \frac{60}{70} - \frac{26}{70} = \frac{17}{35}$

13. $\frac{5}{7} - \frac{4}{9} = \frac{45}{63} - \frac{28}{63} = \frac{17}{63}$
14. $\frac{9}{14} - \frac{10}{21} = \frac{27}{42} - \frac{20}{42} = \frac{1}{6}$

15. $\frac{13}{16} - \frac{7}{24} = \frac{39}{48} - \frac{14}{48} = \frac{25}{48}$
16. $\frac{11}{15} - \frac{31}{75} = \frac{55}{75} - \frac{31}{75} = \frac{8}{25}$

17. $\frac{10}{12} - \frac{3}{42} = \frac{70}{84} - \frac{6}{84} = \frac{16}{21}$
18. $\frac{13}{18} - \frac{23}{45} = \frac{65}{90} - \frac{46}{90} = \frac{19}{90}$

Lesson 8-7   Subtracting fractions with different denominators

Subtract fractions and write the answers in simplest form.

1. $\dfrac{1}{8} - \dfrac{1}{10} = \dfrac{5}{40} - \dfrac{4}{40} = \dfrac{1}{40}$

2. $\dfrac{1}{4} - \dfrac{3}{18} = \dfrac{9}{36} - \dfrac{6}{36} = \dfrac{1}{12}$

3. $\dfrac{1}{2} - \dfrac{3}{7} = \dfrac{7}{14} - \dfrac{6}{14} = \dfrac{1}{14}$

4. $\dfrac{2}{3} - \dfrac{3}{7} = \dfrac{14}{21} - \dfrac{9}{21} = \dfrac{5}{21}$

5. $\dfrac{5}{6} - \dfrac{5}{27} = \dfrac{45}{54} - \dfrac{10}{54} = \dfrac{35}{54}$

6. $\dfrac{11}{12} - \dfrac{13}{21} = \dfrac{77}{84} - \dfrac{52}{84} = \dfrac{25}{84}$

7. $\dfrac{3}{4} - \dfrac{5}{9} = \dfrac{27}{36} - \dfrac{20}{36} = \dfrac{7}{36}$

8. $\dfrac{11}{12} - \dfrac{13}{72} = \dfrac{66}{72} - \dfrac{13}{72} = \dfrac{53}{72}$

9. $\dfrac{1}{4} - \dfrac{4}{17} = \dfrac{17}{68} - \dfrac{16}{68} = \dfrac{1}{68}$

10. $\dfrac{7}{8} - \dfrac{4}{9} = \dfrac{63}{72} - \dfrac{32}{72} = \dfrac{31}{72}$

11. $\dfrac{7}{8} - \dfrac{13}{36} = \dfrac{63}{72} - \dfrac{26}{72} = \dfrac{37}{72}$

12. $\dfrac{13}{18} - \dfrac{17}{30} = \dfrac{65}{90} - \dfrac{51}{90} = \dfrac{7}{45}$

13. $\dfrac{10}{12} - \dfrac{20}{48} = \dfrac{40}{48} - \dfrac{20}{48} = \dfrac{5}{12}$

14. $\dfrac{17}{26} - \dfrac{14}{39} = \dfrac{51}{78} - \dfrac{28}{78} = \dfrac{23}{78}$

15. $\dfrac{7}{9} - \dfrac{35}{72} = \dfrac{56}{72} - \dfrac{35}{72} = \dfrac{7}{24}$

16. $\dfrac{7}{8} - \dfrac{15}{28} = \dfrac{49}{56} - \dfrac{30}{56} = \dfrac{19}{56}$

17. $\dfrac{9}{11} - \dfrac{27}{55} = \dfrac{45}{55} - \dfrac{27}{55} = \dfrac{18}{55}$

18. $\dfrac{9}{16} - \dfrac{11}{24} = \dfrac{27}{48} - \dfrac{22}{48} = \dfrac{5}{48}$

Lesson 8-8   Subtracting fractions with different denominators

Subtract fractions and write the answers in simplest form.

1. $\dfrac{1}{3} - \dfrac{3}{13} = \dfrac{13}{39} - \dfrac{9}{39} = \dfrac{4}{39}$

2. $\dfrac{5}{9} - \dfrac{5}{12} = \dfrac{20}{36} - \dfrac{15}{36} = \dfrac{5}{36}$

3. $\dfrac{1}{2} - \dfrac{4}{15} = \dfrac{15}{30} - \dfrac{8}{30} = \dfrac{7}{30}$

4. $\dfrac{7}{8} - \dfrac{5}{12} = \dfrac{21}{24} - \dfrac{10}{24} = \dfrac{11}{24}$

5. $\dfrac{5}{7} - \dfrac{3}{10} = \dfrac{50}{70} - \dfrac{21}{70} = \dfrac{29}{70}$

6. $\dfrac{2}{3} - \dfrac{1}{8} = \dfrac{16}{24} - \dfrac{3}{24} = \dfrac{13}{24}$

7. $\dfrac{9}{10} - \dfrac{5}{14} = \dfrac{63}{70} - \dfrac{25}{70} = \dfrac{19}{35}$

8. $\dfrac{7}{13} - \dfrac{1}{26} = \dfrac{28}{52} - \dfrac{2}{52} = \dfrac{1}{2}$

9. $\dfrac{3}{4} - \dfrac{5}{12} = \dfrac{9}{12} - \dfrac{5}{12} = \dfrac{1}{3}$

10. $\dfrac{5}{12} - \dfrac{3}{32} = \dfrac{40}{96} - \dfrac{9}{96} = \dfrac{31}{96}$

11. $\dfrac{7}{8} - \dfrac{5}{12} = \dfrac{21}{24} - \dfrac{10}{24} = \dfrac{11}{24}$

12. $\dfrac{7}{9} - \dfrac{19}{27} = \dfrac{21}{27} - \dfrac{19}{27} = \dfrac{2}{27}$

13. $\dfrac{11}{12} - \dfrac{13}{18} = \dfrac{33}{36} - \dfrac{26}{36} = \dfrac{7}{36}$

14. $\dfrac{11}{12} - \dfrac{9}{84} = \dfrac{77}{84} - \dfrac{9}{84} = \dfrac{17}{21}$

15. $\dfrac{13}{14} - \dfrac{27}{42} = \dfrac{39}{42} - \dfrac{27}{42} = \dfrac{2}{7}$

16. $\dfrac{9}{14} - \dfrac{3}{56} = \dfrac{36}{56} - \dfrac{3}{56} = \dfrac{33}{56}$

17. $\dfrac{9}{16} - \dfrac{7}{40} = \dfrac{45}{80} - \dfrac{14}{80} = \dfrac{31}{80}$

18. $\dfrac{13}{18} - \dfrac{11}{27} = \dfrac{39}{54} - \dfrac{22}{54} = \dfrac{17}{54}$

Lesson 8-9   Subtracting fractions with different denominators

Subtract fractions and write the answers in simplest form.

1. $1\dfrac{1}{2} - \dfrac{1}{3} = \dfrac{9}{6} - \dfrac{2}{6} = 1\dfrac{1}{6}$

2. $1\dfrac{1}{4} - \dfrac{5}{6} = \dfrac{15}{12} - \dfrac{10}{12} = \dfrac{5}{12}$

3. $1\dfrac{2}{3} - \dfrac{7}{9} = \dfrac{15}{9} - \dfrac{7}{9} = \dfrac{8}{9}$

4. $1\dfrac{1}{3} - \dfrac{3}{4} = \dfrac{16}{12} - \dfrac{9}{12} = \dfrac{7}{12}$

5. $1\dfrac{1}{6} - \dfrac{7}{12} = \dfrac{14}{12} - \dfrac{7}{12} = \dfrac{7}{12}$

6. $1\dfrac{1}{6} - \dfrac{11}{18} = \dfrac{21}{18} - \dfrac{11}{18} = \dfrac{5}{9}$

7. $1\dfrac{1}{3} - \dfrac{3}{4} = \dfrac{16}{12} - \dfrac{9}{12} = \dfrac{7}{12}$

8. $1\dfrac{2}{5} - \dfrac{13}{20} = \dfrac{28}{20} - \dfrac{13}{20} = \dfrac{3}{4}$

9. $1\dfrac{2}{3} - \dfrac{4}{5} = \dfrac{25}{15} - \dfrac{12}{15} = \dfrac{13}{15}$

10. $1\dfrac{1}{5} - \dfrac{5}{6} = \dfrac{36}{30} - \dfrac{25}{30} = \dfrac{11}{30}$

11. $1\dfrac{3}{8} - \dfrac{11}{12} = \dfrac{33}{24} - \dfrac{22}{24} = \dfrac{11}{24}$

12. $1\dfrac{1}{6} - \dfrac{7}{8} = \dfrac{28}{24} - \dfrac{21}{24} = \dfrac{7}{24}$

13. $1\dfrac{1}{2} - \dfrac{6}{7} = \dfrac{21}{14} - \dfrac{12}{14} = \dfrac{9}{14}$

14. $1\dfrac{3}{7} - \dfrac{8}{9} = \dfrac{90}{63} - \dfrac{56}{63} = \dfrac{34}{63}$

15. $1\dfrac{1}{4} - \dfrac{4}{5} = \dfrac{25}{20} - \dfrac{16}{20} = \dfrac{9}{20}$

16. $1\dfrac{2}{13} - \dfrac{15}{26} = \dfrac{30}{26} - \dfrac{15}{26} = \dfrac{15}{26}$

17. $1\dfrac{1}{6} - \dfrac{3}{5} = \dfrac{35}{30} - \dfrac{18}{30} = \dfrac{17}{30}$

18. $1\dfrac{1}{10} - \dfrac{13}{30} = \dfrac{33}{30} - \dfrac{13}{30} = \dfrac{2}{3}$

Lesson 8-10   Subtracting fractions with different denominators

Subtract fractions and write the answers in simplest form.

1. $1\dfrac{1}{2} - \dfrac{3}{5} = \dfrac{15}{10} - \dfrac{6}{10} = \dfrac{9}{10}$

2. $3\dfrac{1}{2} - 1\dfrac{6}{7} = \dfrac{49}{14} - \dfrac{26}{14} = 1\dfrac{9}{14}$

3. $2\dfrac{1}{4} - 1\dfrac{5}{6} = \dfrac{54}{24} - \dfrac{44}{24} = \dfrac{5}{12}$

4. $2\dfrac{1}{12} - \dfrac{8}{15} = \dfrac{125}{60} - \dfrac{32}{60} = 1\dfrac{11}{20}$

5. $2\dfrac{2}{5} - \dfrac{6}{7} = \dfrac{84}{35} - \dfrac{30}{35} = 1\dfrac{19}{35}$

6. $2\dfrac{1}{6} - 1\dfrac{7}{9} = \dfrac{117}{54} - \dfrac{96}{54} = \dfrac{7}{18}$

7. $1\dfrac{1}{8} - \dfrac{11}{12} = \dfrac{27}{24} - \dfrac{22}{24} = \dfrac{5}{24}$

8. $3\dfrac{1}{4} - \dfrac{5}{7} = \dfrac{91}{28} - \dfrac{20}{28} = 2\dfrac{15}{28}$

9. $2\dfrac{3}{8} - 1\dfrac{7}{9} = \dfrac{171}{72} - \dfrac{128}{72} = \dfrac{43}{72}$

10. $3\dfrac{2}{15} - 2\dfrac{22}{45} = \dfrac{141}{45} - \dfrac{112}{45} = \dfrac{29}{45}$

11. $2\dfrac{1}{14} - \dfrac{15}{28} = \dfrac{58}{28} - \dfrac{15}{28} = 1\dfrac{15}{28}$

12. $2\dfrac{2}{13} - 1\dfrac{16}{65} = \dfrac{140}{65} - \dfrac{81}{65} = \dfrac{59}{65}$

13. $1\dfrac{3}{16} - \dfrac{17}{20} = \dfrac{95}{80} - \dfrac{68}{80} = \dfrac{27}{80}$

14. $3\dfrac{1}{4} - 2\dfrac{9}{10} = \dfrac{65}{20} - \dfrac{58}{20} = \dfrac{7}{20}$

15. $2\dfrac{3}{11} - \dfrac{25}{44} = \dfrac{100}{44} - \dfrac{25}{44} = 1\dfrac{31}{44}$

16. $1\dfrac{5}{12} - \dfrac{17}{18} = \dfrac{51}{36} - \dfrac{34}{36} = \dfrac{17}{36}$

17. $1\dfrac{2}{9} - \dfrac{16}{21} = \dfrac{77}{63} - \dfrac{48}{63} = \dfrac{29}{63}$

18. $2\dfrac{1}{14} - 1\dfrac{22}{35} = \dfrac{145}{70} - \dfrac{114}{70} = \dfrac{31}{70}$

Multiply fractions and write the answers in simplest form.

1. $1\frac{1}{2} \times \frac{1}{3} = \frac{3}{2} \times \frac{1}{3} = \frac{1}{2}$

2. $\frac{1}{3} \times 1\frac{2}{3} = \frac{1}{3} \times \frac{5}{3} = \frac{5}{9}$

3. $1\frac{1}{4} \times \frac{2}{3} = \frac{5}{4} \times \frac{2}{3} = \frac{5}{6}$

4. $\frac{2}{3} \times 1\frac{3}{4} = \frac{2}{3} \times \frac{7}{4} = 1\frac{1}{6}$

5. $1\frac{2}{3} \times 1\frac{1}{5} = \frac{5}{3} \times \frac{6}{5} = 2$

6. $1\frac{2}{3} \times \frac{2}{5} = \frac{5}{3} \times \frac{2}{5} = \frac{2}{3}$

7. $\frac{3}{4} \times 1\frac{2}{5} = \frac{3}{4} \times \frac{7}{5} = \frac{21}{20}$

8. $1\frac{1}{4} \times \frac{3}{4} = \frac{5}{4} \times \frac{3}{4} = \frac{15}{16}$

9. $\frac{2}{7} \times 1\frac{3}{5} = \frac{2}{7} \times \frac{8}{5} = \frac{16}{35}$

10. $1\frac{3}{5} \times 1\frac{2}{7} = \frac{8}{5} \times \frac{9}{7} = 2\frac{2}{35}$

11. $1\frac{2}{3} \times 1\frac{5}{6} = \frac{5}{3} \times \frac{11}{6} = 3\frac{1}{18}$

12. $1\frac{3}{7} \times 1\frac{4}{5} = \frac{10}{7} \times \frac{9}{5} = 2\frac{4}{7}$

13. $1\frac{1}{4} \times 1\frac{2}{3} = \frac{5}{4} \times \frac{5}{3} = 2\frac{1}{12}$

14. $2\frac{2}{9} \times 2\frac{3}{8} = \frac{20}{9} \times \frac{19}{8} = 5\frac{5}{18}$

15. $2\frac{1}{5} \times 1\frac{5}{6} = \frac{11}{5} \times \frac{11}{6} = 4\frac{1}{30}$

16. $1\frac{1}{5} \times 2\frac{5}{8} = \frac{6}{5} \times \frac{21}{8} = 3\frac{3}{20}$

17. $1\frac{3}{4} \times 2\frac{2}{3} = \frac{7}{4} \times \frac{8}{3} = 4\frac{2}{3}$

18. $2\frac{4}{5} \times 1\frac{3}{4} = \frac{14}{5} \times \frac{7}{4} = 4\frac{9}{10}$

Multiply fractions and write the answers in simplest form.

1. $1\frac{2}{3} \times 1\frac{1}{4} = \frac{5}{3} \times \frac{5}{4} = 2\frac{1}{12}$

2. $1\frac{1}{2} \times \frac{5}{7} = \frac{3}{2} \times \frac{5}{7} = 1\frac{1}{14}$

3. $\frac{1}{4} \times 1\frac{2}{5} = \frac{1}{4} \times \frac{7}{5} = \frac{7}{20}$

4. $\frac{2}{3} \times 1\frac{4}{5} = \frac{2}{3} \times \frac{9}{5} = 1\frac{1}{5}$

5. $1\frac{3}{4} \times \frac{1}{6} = \frac{7}{4} \times \frac{1}{6} = \frac{7}{24}$

6. $2\frac{1}{2} \times 2\frac{1}{5} = \frac{5}{2} \times \frac{11}{5} = 5\frac{1}{2}$

7. $2\frac{1}{3} \times 1\frac{3}{5} = \frac{7}{3} \times \frac{8}{5} = 3\frac{11}{15}$

8. $1\frac{2}{3} \times 2\frac{5}{6} = \frac{5}{3} \times \frac{17}{6} = 4\frac{13}{18}$

9. $2\frac{1}{2} \times \frac{3}{7} = \frac{5}{2} \times \frac{3}{7} = 1\frac{1}{14}$

10. $\frac{3}{4} \times 1\frac{4}{5} = \frac{3}{4} \times \frac{9}{5} = 1\frac{7}{20}$

11. $\frac{2}{3} \times 1\frac{5}{8} = \frac{2}{3} \times \frac{13}{8} = 1\frac{1}{12}$

12. $2\frac{5}{6} \times \frac{3}{4} = \frac{17}{6} \times \frac{3}{4} = 2\frac{1}{8}$

13. $1\frac{1}{4} \times \frac{4}{5} = \frac{5}{4} \times \frac{4}{5} = 1$

14. $1\frac{2}{3} \times 1\frac{5}{7} = \frac{5}{3} \times \frac{12}{7} = 2\frac{6}{7}$

15. $\frac{5}{6} \times 2\frac{3}{7} = \frac{5}{6} \times \frac{17}{7} = 2\frac{1}{42}$

16. $2\frac{1}{4} \times 1\frac{6}{7} = \frac{9}{4} \times \frac{13}{7} = 4\frac{5}{28}$

17. $2\frac{3}{4} \times 2\frac{5}{8} = \frac{11}{4} \times \frac{21}{8} = 7\frac{7}{32}$

18. $1\frac{2}{5} \times 2\frac{4}{5} = \frac{7}{5} \times \frac{14}{5} = 3\frac{23}{25}$

Multiply fractions and write the answers in simplest form.

1. $\frac{5}{3} \times 1\frac{2}{3} = \frac{5}{3} \times \frac{5}{3} = 2\frac{7}{9}$

2. $1\frac{1}{4} \times \frac{8}{3} = \frac{5}{4} \times \frac{8}{3} = 3\frac{1}{3}$

3. $\frac{7}{2} \times 1\frac{3}{4} = \frac{7}{2} \times \frac{7}{4} = 6\frac{1}{8}$

4. $\frac{10}{3} \times 1\frac{3}{5} = \frac{10}{3} \times \frac{8}{5} = 5\frac{1}{3}$

5. $1\frac{2}{3} \times \frac{5}{4} = \frac{5}{3} \times \frac{5}{4} = 2\frac{1}{12}$

6. $1\frac{2}{3} \times 1\frac{7}{8} = \frac{5}{3} \times \frac{15}{8} = 3\frac{1}{8}$

7. $1\frac{1}{3} \times \frac{3}{7} = \frac{4}{3} \times \frac{3}{7} = \frac{4}{7}$

8. $2\frac{5}{4} \times 2\frac{2}{5} = \frac{13}{4} \times \frac{12}{5} = 7\frac{4}{5}$

9. $2\frac{2}{5} \times 1\frac{7}{6} = \frac{12}{5} \times \frac{13}{6} = 5\frac{1}{5}$

10. $1\frac{3}{5} \times 2\frac{6}{7} = \frac{8}{5} \times \frac{20}{7} = 4\frac{4}{7}$

11. $1\frac{3}{4} \times 2\frac{5}{8} = \frac{7}{4} \times \frac{21}{8} = 4\frac{19}{32}$

12. $2\frac{9}{8} \times 1\frac{2}{9} = \frac{25}{8} \times \frac{11}{9} = 3\frac{59}{72}$

13. $2\frac{2}{3} \times 2\frac{4}{9} = \frac{8}{3} \times \frac{22}{9} = 6\frac{14}{27}$

14. $1\frac{3}{5} \times 1\frac{2}{9} = \frac{8}{5} \times \frac{11}{9} = 1\frac{43}{45}$

15. $2\frac{5}{4} \times \frac{6}{7} = \frac{13}{4} \times \frac{6}{7} = 2\frac{11}{14}$

16. $2\frac{7}{6} \times 2\frac{5}{7} = \frac{19}{6} \times \frac{19}{7} = 8\frac{25}{42}$

17. $\frac{7}{6} \times 2\frac{5}{8} = \frac{7}{6} \times \frac{21}{8} = 3\frac{1}{16}$

18. $1\frac{3}{4} \times 2\frac{10}{9} = \frac{7}{4} \times \frac{28}{9} = 5\frac{4}{9}$

Multiply fractions and write the answers in simplest form.

1. $2\frac{1}{5} \times \frac{7}{3} = \frac{11}{5} \times \frac{7}{3} = 5\frac{2}{15}$

2. $\frac{3}{4} \times 1\frac{5}{4} = \frac{3}{4} \times \frac{9}{4} = 1\frac{11}{16}$

3. $\frac{1}{4} \times 2\frac{5}{6} = \frac{1}{4} \times \frac{17}{6} = \frac{17}{24}$

4. $1\frac{2}{3} \times 1\frac{5}{8} = \frac{5}{3} \times \frac{13}{8} = 2\frac{17}{24}$

5. $1\frac{2}{3} \times 1\frac{8}{7} = \frac{5}{3} \times \frac{15}{7} = 3\frac{4}{7}$

6. $1\frac{1}{2} \times \frac{5}{9} = \frac{3}{2} \times \frac{5}{9} = \frac{5}{6}$

7. $1\frac{1}{5} \times \frac{3}{5} = \frac{6}{5} \times \frac{3}{5} = \frac{18}{25}$

8. $2\frac{4}{3} \times 1\frac{2}{5} = \frac{10}{3} \times \frac{7}{5} = 4\frac{2}{3}$

9. $\frac{5}{7} \times 1\frac{5}{8} = \frac{5}{7} \times \frac{13}{8} = 1\frac{9}{56}$

10. $2\frac{3}{4} \times \frac{5}{7} = \frac{11}{4} \times \frac{5}{7} = 1\frac{27}{28}$

11. $1\frac{3}{2} \times 1\frac{2}{9} = \frac{5}{2} \times \frac{11}{9} = 3\frac{1}{18}$

12. $\frac{5}{8} \times 2\frac{3}{4} = \frac{5}{8} \times \frac{11}{4} = 1\frac{23}{32}$

13. $\frac{2}{3} \times 2\frac{1}{3} = \frac{2}{3} \times \frac{7}{3} = 1\frac{5}{9}$

14. $1\frac{7}{10} \times 1\frac{3}{8} = \frac{17}{10} \times \frac{11}{8} = 2\frac{27}{80}$

15. $2\frac{4}{5} \times \frac{4}{7} = \frac{14}{5} \times \frac{4}{7} = 1\frac{3}{5}$

16. $2\frac{4}{7} \times 2\frac{7}{8} = \frac{18}{7} \times \frac{23}{8} = 7\frac{11}{28}$

17. $2\frac{2}{9} \times 2\frac{5}{6} = \frac{20}{9} \times \frac{17}{6} = 6\frac{8}{27}$

18. $2\frac{3}{5} \times 1\frac{1}{4} = \frac{13}{5} \times \frac{5}{4} = 3\frac{1}{4}$

Multiply fractions and write the answers in simplest form.

1. $1\frac{2}{5} \times 2\frac{3}{5} = \frac{7}{5} \times \frac{13}{5} = 3\frac{16}{25}$ 2. $2\frac{1}{6} \times 1\frac{2}{5} = \frac{13}{6} \times \frac{7}{5} = 3\frac{1}{30}$

3. $2\frac{1}{3} \times 2\frac{1}{6} = \frac{7}{3} \times \frac{13}{6} = 5\frac{1}{18}$ 4. $3\frac{3}{4} \times 1\frac{4}{5} = \frac{15}{4} \times \frac{9}{5} = 6\frac{3}{4}$

5. $1\frac{3}{4} \times 3\frac{2}{9} = \frac{7}{4} \times \frac{29}{9} = 5\frac{23}{36}$ 6. $1\frac{1}{2} \times 3\frac{5}{8} = \frac{3}{2} \times \frac{29}{8} = 5\frac{7}{16}$

7. $2\frac{1}{6} \times 3\frac{1}{6} = \frac{13}{6} \times \frac{19}{6} = 6\frac{31}{36}$ 8. $3\frac{4}{3} \times 3\frac{5}{4} = \frac{13}{3} \times \frac{17}{4} = 18\frac{5}{12}$

9. $3\frac{4}{3} \times 1\frac{3}{4} = \frac{13}{3} \times \frac{7}{4} = 7\frac{7}{12}$ 10. $2\frac{1}{4} \times 3\frac{7}{9} = \frac{9}{4} \times \frac{34}{9} = 8\frac{1}{2}$

11. $2\frac{4}{5} \times 2\frac{5}{6} = \frac{14}{5} \times \frac{17}{6} = 7\frac{14}{15}$ 12. $2\frac{3}{5} \times 2\frac{9}{10} = \frac{13}{5} \times \frac{29}{10} = 7\frac{27}{50}$

13. $1\frac{1}{2} \times 1\frac{5}{6} = \frac{3}{2} \times \frac{11}{6} = 2\frac{3}{4}$ 14. $1\frac{7}{6} \times 2\frac{8}{7} = \frac{13}{6} \times \frac{22}{7} = 6\frac{17}{21}$

15. $3\frac{1}{2} \times 3\frac{4}{5} = \frac{7}{2} \times \frac{19}{5} = 13\frac{3}{10}$ 16. $3\frac{3}{4} \times 3\frac{5}{6} = \frac{15}{4} \times \frac{23}{6} = 14\frac{3}{8}$

17. $2\frac{2}{3} \times 1\frac{4}{7} = \frac{8}{3} \times \frac{11}{7} = 4\frac{4}{21}$ 18. $1\frac{11}{10} \times 2\frac{5}{6} = \frac{21}{10} \times \frac{17}{6} = 5\frac{19}{20}$

Multiply fractions and write the answers in simplest form.

1. $2\frac{1}{3} \times 2\frac{1}{4} = \frac{7}{3} \times \frac{9}{4} = 5\frac{1}{4}$ 2. $2\frac{1}{2} \times 2\frac{1}{4} = \frac{5}{2} \times \frac{9}{4} = 5\frac{5}{8}$

3. $3\frac{2}{3} \times 1\frac{1}{5} = \frac{11}{3} \times \frac{6}{5} = 4\frac{2}{5}$ 4. $3\frac{2}{3} \times 1\frac{1}{3} = \frac{11}{3} \times \frac{4}{3} = 4\frac{8}{9}$

5. $1\frac{3}{4} \times 2\frac{2}{5} = \frac{7}{4} \times \frac{12}{5} = 4\frac{1}{5}$ 6. $1\frac{3}{4} \times 2\frac{5}{6} = \frac{7}{4} \times \frac{17}{6} = 4\frac{23}{24}$

7. $2\frac{1}{4} \times 2\frac{5}{6} = \frac{9}{4} \times \frac{17}{6} = 6\frac{3}{8}$ 8. $2\frac{1}{6} \times 3\frac{7}{12} = \frac{13}{6} \times \frac{43}{12} = 7\frac{55}{72}$

9. $3\frac{3}{5} \times 2\frac{3}{10} = \frac{18}{5} \times \frac{23}{10} = 8\frac{7}{35}$ 10. $3\frac{2}{5} \times 3\frac{1}{6} = \frac{17}{5} \times \frac{19}{6} = 10\frac{23}{30}$

11. $3\frac{2}{5} \times 3\frac{5}{6} = \frac{17}{5} \times \frac{23}{6} = 13\frac{1}{30}$ 12. $3\frac{5}{6} \times 2\frac{4}{9} = \frac{23}{6} \times \frac{22}{9} = 9\frac{10}{27}$

13. $1\frac{3}{4} \times 3\frac{9}{10} = \frac{7}{4} \times \frac{39}{10} = 6\frac{33}{40}$ 14. $1\frac{3}{7} \times 1\frac{4}{5} = \frac{10}{7} \times \frac{9}{5} = 2\frac{4}{7}$

15. $2\frac{2}{3} \times 2\frac{7}{9} = \frac{8}{3} \times \frac{25}{9} = 7\frac{11}{27}$ 16. $2\frac{1}{6} \times 1\frac{6}{7} = \frac{13}{6} \times \frac{13}{7} = 4\frac{1}{42}$

17. $3\frac{3}{4} \times 1\frac{9}{10} = \frac{15}{4} \times \frac{19}{10} = 7\frac{1}{8}$ 18. $3\frac{5}{6} \times 3\frac{7}{9} = \frac{23}{6} \times \frac{34}{6} = 21\frac{13}{18}$

Multiply fractions and write the answers in simplest form.

1. $4\frac{1}{2} \times 3\frac{2}{3} = \frac{9}{2} \times \frac{11}{3} = 16\frac{1}{2}$ 2. $3\frac{1}{2} \times 3\frac{1}{2} = \frac{7}{2} \times \frac{7}{2} = 12\frac{1}{4}$

3. $3\frac{1}{3} \times 4\frac{2}{5} = \frac{10}{3} \times \frac{22}{5} = 14\frac{2}{3}$ 4. $4\frac{2}{3} \times 2\frac{1}{2} = \frac{14}{3} \times \frac{5}{2} = 11\frac{2}{3}$

5. $3\frac{2}{3} \times 2\frac{3}{4} = \frac{11}{3} \times \frac{11}{4} = 10\frac{1}{12}$ 6. $1\frac{3}{4} \times 3\frac{2}{3} = \frac{7}{4} \times \frac{11}{3} = 6\frac{5}{12}$

7. $4\frac{1}{6} \times 2\frac{2}{3} = \frac{25}{6} \times \frac{8}{3} = 11\frac{1}{9}$ 8. $2\frac{1}{5} \times 2\frac{2}{5} = \frac{11}{5} \times \frac{12}{5} = 5\frac{7}{25}$

9. $3\frac{3}{5} \times 4\frac{1}{2} = \frac{18}{5} \times \frac{9}{2} = 16\frac{1}{5}$ 10. $2\frac{3}{4} \times 4\frac{3}{5} = \frac{11}{4} \times \frac{23}{5} = 12\frac{13}{20}$

11. $4\frac{1}{5} \times 1\frac{3}{4} = \frac{21}{5} \times \frac{7}{4} = 7\frac{7}{20}$ 12. $3\frac{1}{3} \times 1\frac{1}{6} = \frac{10}{3} \times \frac{7}{6} = 3\frac{8}{9}$

13. $2\frac{2}{3} \times 3\frac{1}{3} = \frac{8}{3} \times \frac{10}{3} = 8\frac{8}{9}$ 14. $1\frac{3}{5} \times 3\frac{1}{3} = \frac{8}{5} \times \frac{10}{3} = 5\frac{1}{3}$

15. $4\frac{2}{5} \times 2\frac{3}{5} = \frac{22}{5} \times \frac{13}{5} = 11\frac{1}{5}$ 16. $2\frac{5}{6} \times 2\frac{3}{4} = \frac{17}{6} \times \frac{11}{4} = 7\frac{19}{24}$

17. $3\frac{1}{4} \times 3\frac{2}{5} = \frac{13}{4} \times \frac{17}{5} = 11\frac{1}{20}$ 18. $4\frac{1}{3} \times 3\frac{4}{5} = \frac{13}{3} \times \frac{19}{5} = 16\frac{7}{15}$

Multiply fractions and write the answers in simplest form.

1. $4\frac{1}{2} \times 2\frac{2}{3} = \frac{9}{2} \times \frac{8}{3} = 12$ 2. $5\frac{1}{3} \times 1\frac{3}{2} = \frac{16}{3} \times \frac{5}{2} = 13\frac{1}{3}$

3. $3\frac{3}{4} \times 4\frac{1}{5} = \frac{15}{4} \times \frac{21}{5} = 15\frac{3}{4}$ 4. $3\frac{3}{4} \times 3\frac{2}{3} = \frac{15}{4} \times \frac{11}{3} = 13\frac{3}{4}$

5. $5\frac{2}{3} \times 1\frac{2}{3} = \frac{17}{3} \times \frac{5}{3} = 9\frac{4}{9}$ 6. $2\frac{4}{5} \times 5\frac{1}{2} = \frac{14}{5} \times \frac{11}{2} = 15\frac{2}{5}$

7. $2\frac{2}{5} \times 3\frac{1}{6} = \frac{12}{5} \times \frac{19}{6} = 7\frac{3}{5}$ 8. $2\frac{1}{2} \times 3\frac{7}{6} = \frac{5}{2} \times \frac{25}{6} = 10\frac{5}{12}$

9. $3\frac{2}{3} \times 1\frac{8}{9} = \frac{11}{3} \times \frac{17}{9} = 6\frac{25}{27}$ 10. $1\frac{3}{4} \times 5\frac{4}{3} = \frac{7}{4} \times \frac{19}{3} = 11\frac{1}{12}$

11. $2\frac{3}{4} \times 5\frac{3}{8} = \frac{11}{4} \times \frac{43}{8} = 14\frac{25}{32}$ 12. $5\frac{2}{3} \times 2\frac{7}{8} = \frac{17}{3} \times \frac{23}{8} = 16\frac{7}{24}$

13. $5\frac{1}{3} \times 3\frac{2}{5} = \frac{16}{3} \times \frac{17}{5} = 18\frac{2}{15}$ 14. $4\frac{1}{4} \times 3\frac{2}{5} = \frac{17}{4} \times \frac{17}{5} = 14\frac{9}{20}$

15. $4\frac{1}{3} \times 3\frac{3}{7} = \frac{13}{3} \times \frac{24}{7} = 14\frac{6}{7}$ 16. $3\frac{2}{5} \times 2\frac{5}{6} = \frac{17}{5} \times \frac{17}{6} = 9\frac{19}{30}$

17. $2\frac{5}{6} \times 2\frac{5}{8} = \frac{17}{6} \times \frac{21}{8} = 7\frac{7}{16}$ 18. $4\frac{3}{4} \times 4\frac{2}{3} = \frac{19}{4} \times \frac{14}{3} = 22\frac{1}{6}$

Multiply fractions and write the answers in simplest form.

1. $6\frac{1}{2} \times 7\frac{1}{2} = \frac{13}{2} \times \frac{15}{2} = 48\frac{3}{4}$ 2. $6\frac{1}{2} \times 2\frac{1}{3} = \frac{13}{2} \times \frac{7}{3} = 15\frac{1}{6}$

3. $3\frac{2}{3} \times 5\frac{1}{3} = \frac{11}{3} \times \frac{16}{3} = 19\frac{5}{9}$ 4. $5\frac{2}{3} \times 5\frac{1}{3} = \frac{17}{3} \times \frac{16}{3} = 30\frac{2}{9}$

5. $5\frac{2}{3} \times 6\frac{1}{4} = \frac{17}{3} \times \frac{25}{4} = 35\frac{5}{12}$ 6. $4\frac{1}{4} \times 4\frac{2}{3} = \frac{17}{4} \times \frac{14}{3} = 19\frac{5}{6}$

7. $5\frac{1}{3} \times 3\frac{2}{3} = \frac{16}{3} \times \frac{11}{3} = 19\frac{5}{9}$ 8. $5\frac{1}{2} \times 3\frac{3}{4} = \frac{11}{2} \times \frac{15}{4} = 20\frac{5}{8}$

9. $4\frac{1}{2} \times 6\frac{1}{3} = \frac{9}{2} \times \frac{19}{3} = 28\frac{1}{2}$ 10. $2\frac{4}{5} \times 6\frac{1}{3} = \frac{14}{5} \times \frac{19}{3} = 17\frac{11}{15}$

11. $6\frac{3}{4} \times 6\frac{1}{2} = \frac{27}{4} \times \frac{13}{2} = 43\frac{7}{8}$ 12. $5\frac{1}{3} \times 4\frac{3}{4} = \frac{16}{3} \times \frac{19}{4} = 25\frac{1}{3}$

13. $7\frac{1}{2} \times 6\frac{1}{3} = \frac{15}{2} \times \frac{19}{3} = 47\frac{1}{2}$ 14. $3\frac{4}{5} \times 5\frac{2}{3} = \frac{19}{5} \times \frac{17}{3} = 21\frac{8}{15}$

15. $6\frac{2}{3} \times 5\frac{1}{2} = \frac{20}{3} \times \frac{11}{2} = 36\frac{2}{3}$ 16. $4\frac{2}{3} \times 3\frac{2}{3} = \frac{14}{3} \times \frac{11}{3} = 17\frac{1}{9}$

17. $6\frac{1}{4} \times 4\frac{3}{4} = \frac{25}{4} \times \frac{19}{4} = 29\frac{11}{16}$ 18. $6\frac{1}{2} \times 6\frac{3}{4} = \frac{13}{2} \times \frac{27}{4} = 43\frac{7}{8}$

Multiply fractions and write the answers in simplest form.

1. $7\frac{1}{2} \times 4\frac{1}{3} = \frac{15}{2} \times \frac{13}{3} = 32\frac{1}{2}$ 2. $8\frac{1}{2} \times 5\frac{1}{2} = \frac{17}{2} \times \frac{11}{2} = 46\frac{3}{4}$

3. $8\frac{1}{3} \times 5\frac{2}{3} = \frac{25}{3} \times \frac{17}{3} = 47\frac{2}{9}$ 4. $5\frac{1}{3} \times 7\frac{1}{2} = \frac{16}{3} \times \frac{15}{2} = 40$

5. $4\frac{1}{2} \times 6\frac{1}{2} = \frac{9}{2} \times \frac{13}{2} = 29\frac{1}{4}$ 6. $3\frac{2}{3} \times 5\frac{1}{3} = \frac{11}{3} \times \frac{16}{3} = 19\frac{5}{9}$

7. $6\frac{2}{3} \times 5\frac{1}{3} = \frac{20}{3} \times \frac{16}{3} = 35\frac{5}{9}$ 8. $4\frac{1}{2} \times 6\frac{1}{2} = \frac{9}{2} \times \frac{13}{2} = 29\frac{1}{4}$

9. $6\frac{1}{2} \times 2\frac{1}{2} = \frac{13}{2} \times \frac{5}{2} = 16\frac{1}{4}$ 10. $7\frac{2}{3} \times 5\frac{1}{3} = \frac{23}{3} \times \frac{16}{3} = 40\frac{8}{9}$

11. $5\frac{2}{3} \times 6\frac{1}{2} = \frac{17}{3} \times \frac{13}{2} = 36\frac{5}{6}$ 12. $5\frac{1}{2} \times 5\frac{1}{2} = \frac{11}{2} \times \frac{11}{2} = 30\frac{1}{4}$

13. $6\frac{1}{2} \times 6\frac{2}{3} = \frac{13}{2} \times \frac{20}{3} = 43\frac{1}{3}$ 14. $6\frac{2}{3} \times 3\frac{1}{2} = \frac{20}{3} \times \frac{7}{2} = 23\frac{1}{3}$

15. $7\frac{1}{3} \times 3\frac{2}{3} = \frac{22}{3} \times \frac{11}{3} = 26\frac{8}{9}$ 16. $8\frac{1}{3} \times 9\frac{2}{3} = \frac{25}{3} \times \frac{29}{3} = 80\frac{5}{9}$

17. $7\frac{1}{2} \times 5\frac{1}{2} = \frac{15}{2} \times \frac{11}{2} = 41\frac{1}{4}$ 18. $7\frac{2}{3} \times 7\frac{1}{2} = \frac{23}{3} \times \frac{15}{2} = 57\frac{1}{2}$

Divide fractions and write the answers in simplest form.

1. $2 \div \frac{1}{3} = 2 \times \frac{3}{1} = 6$ 2. $\frac{1}{2} \div 2 = \frac{1}{2} \times \frac{1}{2} = \frac{1}{4}$

3. $2 \div \frac{3}{2} = 2 \times \frac{2}{3} = 1\frac{1}{3}$ 4. $\frac{1}{2} \div 3 = \frac{1}{2} \times \frac{1}{3} = \frac{1}{6}$

5. $2 \div \frac{1}{4} = 2 \times \frac{4}{1} = 8$ 6. $\frac{1}{2} \div 4 = \frac{1}{2} \times \frac{1}{4} = \frac{1}{8}$

7. $2 \div \frac{2}{5} = 2 \times \frac{5}{2} = 5$ 8. $\frac{1}{3} \div 2 = \frac{1}{3} \times \frac{1}{2} = \frac{1}{6}$

9. $3 \div \frac{3}{5} = 3 \times \frac{5}{3} = 5$ 10. $\frac{2}{3} \div 3 = \frac{2}{3} \times \frac{1}{3} = \frac{2}{9}$

11. $3 \div \frac{3}{4} = 3 \times \frac{4}{3} = 4$ 12. $\frac{1}{4} \div 2 = \frac{1}{4} \times \frac{1}{2} = \frac{1}{8}$

13. $3 \div \frac{4}{5} = 3 \times \frac{5}{4} = 3\frac{3}{4}$ 14. $\frac{3}{4} \div 3 = \frac{3}{4} \times \frac{1}{3} = \frac{1}{4}$

15. $4 \div \frac{1}{6} = 4 \times \frac{6}{1} = 24$ 16. $\frac{1}{5} \div 2 = \frac{1}{5} \times \frac{1}{2} = \frac{1}{10}$

17. $4 \div \frac{5}{6} = 4 \times \frac{6}{5} = 4\frac{4}{5}$ 18. $\frac{2}{5} \div 3 = \frac{2}{5} \times \frac{1}{3} = \frac{2}{15}$

Divide fractions and write the answers in simplest form.

1. $2 \div 2\frac{1}{2} = 2 \times \frac{2}{5} = \frac{4}{5}$ 2. $1\frac{1}{2} \div 2 = \frac{3}{2} \times \frac{1}{2} = \frac{3}{4}$

3. $3 \div 2\frac{1}{3} = 3 \times \frac{3}{7} = 1\frac{2}{7}$ 4. $1\frac{1}{3} \div 2 = \frac{4}{3} \times \frac{1}{2} = \frac{2}{3}$

5. $2 \div 1\frac{2}{3} = 2 \times \frac{3}{5} = 1\frac{1}{5}$ 6. $2\frac{2}{3} \div 2 = \frac{8}{3} \times \frac{1}{2} = 1\frac{1}{3}$

7. $1 \div 1\frac{1}{4} = 1 \times \frac{4}{5} = \frac{4}{5}$ 8. $2\frac{1}{5} \div 3 = \frac{11}{5} \times \frac{1}{3} = \frac{11}{15}$

9. $3 \div 1\frac{3}{4} = 3 \times \frac{4}{7} = 1\frac{5}{7}$ 10. $1\frac{1}{4} \div 3 = \frac{5}{4} \times \frac{1}{3} = \frac{5}{12}$

11. $1 \div 2\frac{1}{5} = 1 \times \frac{5}{11} = \frac{5}{11}$ 12. $2\frac{3}{4} \div 3 = \frac{11}{4} \times \frac{1}{3} = \frac{11}{12}$

13. $2 \div 2\frac{2}{5} = 2 \times \frac{5}{12} = \frac{5}{6}$ 14. $1\frac{2}{5} \div 2 = \frac{7}{5} \times \frac{1}{2} = \frac{7}{10}$

15. $3 \div 1\frac{3}{5} = 3 \times \frac{5}{8} = 1\frac{7}{8}$ 16. $1\frac{1}{6} \div 3 = \frac{7}{6} \times \frac{1}{3} = \frac{7}{18}$

17. $2 \div 1\frac{4}{5} = 2 \times \frac{5}{9} = 1\frac{1}{9}$ 18. $2\frac{3}{5} \div 3 = \frac{13}{5} \times \frac{1}{3} = \frac{13}{15}$

Divide fractions and write the answers in simplest form.

1. $\frac{1}{2} \div 1\frac{1}{2} = \frac{1}{2} \times \frac{2}{3} = \frac{1}{3}$

2. $1\frac{1}{2} \div \frac{2}{3} = \frac{3}{2} \times \frac{3}{2} = \frac{9}{4}$

3. $\frac{3}{4} \div \frac{5}{6} = \frac{3}{4} \times \frac{6}{5} = \frac{9}{10}$

4. $\frac{1}{4} \div 1\frac{3}{4} = \frac{1}{4} \times \frac{4}{7} = \frac{1}{7}$

5. $\frac{2}{3} \div 1\frac{1}{5} = \frac{2}{3} \times \frac{5}{6} = \frac{5}{9}$

6. $1\frac{1}{3} \div \frac{2}{5} = \frac{4}{3} \times \frac{5}{2} = 3\frac{1}{3}$

7. $\frac{2}{3} \div \frac{3}{4} = \frac{2}{3} \times \frac{4}{3} = \frac{8}{9}$

8. $1\frac{2}{3} \div \frac{3}{5} = \frac{5}{3} \times \frac{5}{3} = 2\frac{7}{9}$

9. $\frac{2}{3} \div 1\frac{1}{3} = \frac{2}{3} \times \frac{3}{4} = \frac{1}{2}$

10. $\frac{2}{5} \div 1\frac{1}{2} = \frac{2}{5} \times \frac{2}{3} = \frac{4}{15}$

11. $\frac{4}{5} \div \frac{5}{6} = \frac{4}{5} \times \frac{6}{5} = \frac{24}{25}$

12. $1\frac{1}{5} \div 2\frac{1}{3} = \frac{6}{5} \times \frac{3}{7} = \frac{18}{35}$

13. $\frac{5}{6} \div 1\frac{1}{11} = \frac{5}{6} \times \frac{11}{12} = \frac{55}{72}$

14. $\frac{3}{4} \div 1\frac{2}{3} = \frac{3}{4} \times \frac{3}{5} = \frac{9}{20}$

15. $\frac{4}{5} \div 1\frac{1}{5} = \frac{4}{5} \times \frac{5}{6} = \frac{2}{3}$

16. $2\frac{3}{7} \div \frac{3}{4} = \frac{17}{7} \times \frac{4}{3} = 3\frac{5}{21}$

17. $\frac{7}{9} \div 1\frac{2}{5} = \frac{7}{9} \times \frac{5}{7} = \frac{5}{9}$

18. $\frac{4}{5} \div 1\frac{1}{3} = \frac{4}{5} \times \frac{3}{4} = \frac{3}{5}$

Divide fractions and write the answers in simplest form.

1. $\frac{1}{2} \div \frac{7}{6} = \frac{1}{2} \times \frac{6}{7} = \frac{3}{7}$

2. $\frac{4}{5} \div \frac{8}{5} = \frac{4}{5} \times \frac{5}{8} = \frac{1}{2}$

3. $\frac{3}{4} \div \frac{10}{9} = \frac{3}{4} \times \frac{9}{10} = \frac{27}{40}$

4. $\frac{9}{14} \div \frac{5}{8} = \frac{9}{14} \times \frac{8}{5} = 1\frac{1}{35}$

5. $\frac{11}{12} \div \frac{15}{14} = \frac{11}{12} \times \frac{14}{15} = \frac{77}{90}$

6. $\frac{9}{10} \div \frac{9}{20} = \frac{9}{10} \times \frac{20}{9} = 2$

7. $\frac{2}{3} \div \frac{7}{6} = \frac{2}{3} \times \frac{6}{7} = \frac{4}{7}$

8. $\frac{5}{6} \div \frac{9}{10} = \frac{5}{6} \times \frac{10}{9} = \frac{25}{27}$

9. $\frac{5}{6} \div \frac{9}{8} = \frac{5}{6} \times \frac{8}{9} = \frac{20}{27}$

10. $\frac{8}{9} \div \frac{18}{17} = \frac{8}{9} \times \frac{17}{18} = \frac{68}{81}$

11. $\frac{11}{12} \div \frac{11}{10} = \frac{11}{12} \times \frac{10}{11} = \frac{5}{6}$

12. $\frac{7}{8} \div \frac{12}{11} = \frac{7}{8} \times \frac{11}{12} = \frac{77}{96}$

13. $\frac{3}{4} \div \frac{7}{8} = \frac{3}{4} \times \frac{8}{7} = \frac{6}{7}$

14. $\frac{4}{5} \div \frac{8}{7} = \frac{4}{5} \times \frac{7}{8} = \frac{7}{10}$

15. $\frac{5}{8} \div \frac{9}{16} = \frac{5}{8} \times \frac{16}{9} = 1\frac{1}{9}$

16. $\frac{5}{6} \div \frac{15}{14} = \frac{5}{6} \times \frac{14}{15} = \frac{7}{9}$

17. $\frac{11}{15} \div \frac{7}{30} = \frac{11}{15} \times \frac{30}{7} = 3\frac{1}{7}$

18. $\frac{5}{8} \div \frac{16}{5} = \frac{5}{8} \times \frac{5}{16} = \frac{25}{128}$

Divide fractions and write the answers in simplest form.

1. $\frac{1}{2} \div \frac{2}{3} = \frac{1}{2} \times \frac{3}{2} = \frac{3}{4}$

2. $\frac{5}{3} \div \frac{15}{7} = \frac{5}{3} \times \frac{7}{15} = \frac{7}{9}$

3. $\frac{2}{3} \div \frac{5}{6} = \frac{2}{3} \times \frac{6}{5} = \frac{4}{5}$

4. $\frac{3}{2} \div \frac{5}{4} = \frac{3}{2} \times \frac{4}{5} = 1\frac{1}{5}$

5. $\frac{1}{4} \div \frac{12}{7} = \frac{1}{4} \times \frac{7}{12} = \frac{7}{48}$

6. $\frac{4}{3} \div \frac{3}{4} = \frac{4}{3} \times \frac{4}{3} = \frac{16}{9}$

7. $\frac{1}{3} \div \frac{9}{5} = \frac{1}{3} \times \frac{5}{9} = \frac{5}{27}$

8. $\frac{5}{2} \div \frac{8}{3} = \frac{5}{2} \times \frac{3}{8} = \frac{15}{16}$

9. $\frac{3}{4} \div \frac{8}{9} = \frac{3}{4} \times \frac{9}{8} = \frac{27}{32}$

10. $\frac{7}{4} \div \frac{9}{14} = \frac{7}{4} \times \frac{14}{9} = 2\frac{13}{18}$

11. $\frac{5}{6} \div \frac{12}{5} = \frac{5}{6} \times \frac{5}{12} = \frac{25}{72}$

12. $\frac{4}{5} \div \frac{3}{10} = \frac{4}{5} \times \frac{10}{3} = 2\frac{2}{3}$

13. $\frac{2}{5} \div \frac{15}{4} = \frac{2}{5} \times \frac{4}{15} = \frac{8}{75}$

14. $\frac{9}{4} \div \frac{7}{2} = \frac{9}{4} \times \frac{2}{7} = \frac{9}{14}$

15. $\frac{3}{7} \div \frac{14}{5} = \frac{3}{7} \times \frac{5}{14} = \frac{15}{98}$

16. $\frac{10}{3} \div \frac{6}{5} = \frac{10}{3} \times \frac{5}{6} = 2\frac{7}{9}$

17. $\frac{4}{5} \div \frac{8}{3} = \frac{4}{5} \times \frac{3}{8} = \frac{3}{10}$

18. $\frac{11}{7} \div \frac{5}{14} = \frac{7}{11} \times \frac{14}{5} = 1\frac{43}{55}$

Divide fractions and write the answers in simplest form.

1. $1\frac{1}{3} \div \frac{2}{3} = \frac{4}{3} \times \frac{3}{2} = 2$

2. $2\frac{1}{3} \div 1\frac{1}{2} = \frac{7}{3} \times \frac{2}{3} = 1\frac{5}{9}$

3. $1\frac{1}{2} \div 1\frac{1}{3} = \frac{3}{2} \times \frac{3}{4} = 1\frac{1}{8}$

4. $1\frac{3}{4} \div \frac{2}{3} = \frac{7}{4} \times \frac{3}{2} = 2\frac{5}{8}$

5. $\frac{3}{4} \div 1\frac{2}{3} = \frac{3}{4} \times \frac{3}{5} = \frac{9}{20}$

6. $\frac{1}{2} \div 2\frac{1}{4} = \frac{1}{2} \times \frac{4}{9} = \frac{2}{9}$

7. $1\frac{1}{5} \div 2\frac{1}{2} = \frac{6}{5} \times \frac{2}{5} = \frac{12}{25}$

8. $2\frac{3}{5} \div \frac{5}{2} = \frac{13}{5} \times \frac{2}{5} = 1\frac{1}{25}$

9. $2\frac{1}{4} \div 1\frac{2}{3} = \frac{9}{4} \times \frac{3}{5} = 1\frac{7}{20}$

10. $\frac{2}{5} \div \frac{3}{4} = \frac{2}{5} \times \frac{4}{3} = \frac{8}{15}$

11. $1\frac{2}{5} \div 1\frac{1}{4} = \frac{7}{5} \times \frac{4}{5} = 1\frac{3}{25}$

12. $1\frac{1}{2} \div 1\frac{2}{5} = \frac{3}{2} \times \frac{5}{7} = 1\frac{1}{14}$

13. $\frac{3}{7} \div 1\frac{1}{6} = \frac{3}{7} \times \frac{6}{7} = \frac{18}{49}$

14. $2\frac{4}{5} \div 2\frac{2}{5} = \frac{14}{5} \times \frac{5}{12} = 1\frac{1}{6}$

15. $2\frac{3}{5} \div 2\frac{1}{3} = \frac{13}{5} \times \frac{3}{7} = 1\frac{4}{35}$

16. $2\frac{1}{3} \div \frac{5}{6} = \frac{7}{3} \times \frac{6}{5} = 2\frac{4}{5}$

17. $1\frac{5}{6} \div \frac{11}{7} = \frac{11}{6} \times \frac{7}{11} = 1\frac{1}{6}$

18. $\frac{3}{4} \div 1\frac{1}{2} = \frac{3}{4} \times \frac{2}{3} = \frac{1}{2}$

Divide fractions and write the answers in simplest form.

1. $2\frac{1}{2} \div 2\frac{1}{4} = \frac{5}{2} \times \frac{4}{9} = 1\frac{1}{9}$
2. $1\frac{3}{5} \div 2\frac{1}{3} = \frac{8}{5} \times \frac{3}{7} = \frac{24}{35}$
3. $3\frac{2}{3} \div \frac{2}{3} = \frac{11}{3} \times \frac{3}{2} = 5\frac{1}{2}$
4. $1\frac{2}{3} \div \frac{2}{5} = \frac{5}{3} \times \frac{5}{2} = 4\frac{1}{6}$
5. $1\frac{3}{5} \div 3\frac{1}{6} = \frac{8}{5} \times \frac{6}{19} = \frac{48}{95}$
6. $3\frac{1}{4} \div 2\frac{1}{3} = \frac{13}{4} \times \frac{3}{7} = 1\frac{11}{28}$
7. $2\frac{1}{3} \div 2\frac{2}{5} = \frac{7}{3} \times \frac{5}{12} = \frac{35}{36}$
8. $1\frac{2}{5} \div 1\frac{3}{4} = \frac{7}{5} \times \frac{4}{7} = \frac{4}{5}$
9. $3\frac{1}{2} \div 2\frac{3}{4} = \frac{7}{2} \times \frac{4}{11} = 1\frac{3}{11}$
10. $3\frac{1}{2} \div 1\frac{2}{3} = \frac{7}{2} \times \frac{3}{5} = 2\frac{1}{10}$
11. $\frac{1}{4} \div \frac{2}{5} = \frac{1}{4} \times \frac{5}{2} = \frac{5}{8}$
12. $2\frac{3}{4} \div 3\frac{1}{4} = \frac{11}{4} \times \frac{4}{13} = \frac{11}{13}$
13. $\frac{2}{7} \div 2\frac{1}{3} = \frac{2}{7} \times \frac{3}{7} = \frac{6}{49}$
14. $3\frac{4}{5} \div \frac{2}{5} = \frac{19}{5} \times \frac{5}{2} = 9\frac{1}{2}$
15. $2\frac{4}{5} \div 3\frac{1}{2} = \frac{14}{5} \times \frac{2}{7} = \frac{4}{5}$
16. $3\frac{3}{7} \div 3\frac{1}{2} = \frac{24}{7} \times \frac{2}{7} = \frac{48}{49}$
17. $3\frac{3}{7} \div 3\frac{2}{3} = \frac{24}{7} \times \frac{3}{11} = \frac{72}{77}$
18. $\frac{5}{6} \div 3\frac{1}{2} = \frac{5}{6} \times \frac{2}{7} = \frac{5}{21}$

Divide fractions and write the answers in simplest form.

1. $3\frac{1}{2} \div \frac{2}{3} = \frac{7}{2} \times \frac{3}{2} = 5\frac{1}{4}$
2. $3\frac{3}{4} \div 1\frac{1}{3} = \frac{15}{4} \times \frac{3}{4} = 2\frac{13}{16}$
3. $4\frac{1}{3} \div 2\frac{1}{2} = \frac{13}{3} \times \frac{2}{5} = 1\frac{11}{15}$
4. $2\frac{2}{5} \div 2\frac{1}{5} = \frac{12}{5} \times \frac{5}{11} = 1\frac{1}{11}$
5. $5\frac{2}{3} \div 2\frac{1}{3} = \frac{17}{3} \times \frac{3}{7} = 2\frac{3}{7}$
6. $4\frac{2}{3} \div 3\frac{1}{2} = \frac{14}{3} \times \frac{2}{7} = 1\frac{1}{3}$
7. $3\frac{1}{4} \div 2\frac{3}{4} = \frac{13}{4} \times \frac{4}{11} = 1\frac{2}{11}$
8. $5\frac{1}{6} \div 2\frac{1}{3} = \frac{31}{6} \times \frac{3}{7} = 2\frac{3}{14}$
9. $5\frac{3}{5} \div 3\frac{1}{3} = \frac{28}{5} \times \frac{3}{10} = 1\frac{17}{25}$
10. $3\frac{4}{5} \div 2\frac{2}{3} = \frac{19}{5} \times \frac{3}{8} = 1\frac{3}{40}$
11. $2\frac{1}{6} \div 1\frac{2}{3} = \frac{13}{6} \times \frac{3}{5} = 1\frac{3}{10}$
12. $5\frac{3}{7} \div 3\frac{2}{5} = \frac{38}{7} \times \frac{5}{17} = 1\frac{71}{119}$
13. $4\frac{2}{5} \div 3\frac{1}{4} = \frac{22}{5} \times \frac{4}{13} = 1\frac{23}{65}$
14. $4\frac{1}{4} \div 1\frac{1}{2} = \frac{17}{4} \times \frac{2}{3} = 1\frac{5}{6}$
15. $4\frac{3}{4} \div 1\frac{2}{3} = \frac{19}{4} \times \frac{3}{5} = 2\frac{17}{20}$
16. $3\frac{2}{3} \div 3\frac{1}{4} = \frac{11}{3} \times \frac{4}{13} = 1\frac{5}{39}$
17. $5\frac{5}{6} \div 4\frac{1}{6} = \frac{35}{6} \times \frac{6}{25} = 1\frac{2}{5}$
18. $5\frac{3}{4} \div 3\frac{2}{5} = \frac{23}{4} \times \frac{5}{17} = 1\frac{47}{68}$

Divide fractions and write the answers in simplest form.

1. $3\frac{1}{2} \div 4\frac{2}{3} = \frac{7}{2} \times \frac{3}{14} = \frac{3}{4}$
2. $2\frac{1}{2} \div 3\frac{3}{4} = \frac{5}{2} \times \frac{4}{15} = \frac{2}{3}$
3. $5\frac{1}{3} \div 2\frac{2}{3} = \frac{16}{3} \times \frac{3}{8} = 2$
4. $1\frac{2}{3} \div 4\frac{1}{2} = \frac{5}{3} \times \frac{2}{9} = \frac{10}{27}$
5. $2\frac{2}{3} \div 3\frac{1}{2} = \frac{8}{3} \times \frac{2}{7} = \frac{16}{21}$
6. $2\frac{1}{4} \div 4\frac{1}{3} = \frac{9}{4} \times \frac{3}{13} = \frac{27}{52}$
7. $2\frac{1}{5} \div 6\frac{1}{2} = \frac{11}{5} \times \frac{2}{13} = \frac{22}{65}$
8. $1\frac{1}{3} \div 3\frac{1}{2} = \frac{4}{3} \times \frac{2}{7} = \frac{8}{21}$
9. $1\frac{3}{5} \div 2\frac{2}{3} = \frac{8}{5} \times \frac{3}{8} = \frac{3}{5}$
10. $3\frac{3}{4} \div 4\frac{1}{3} = \frac{15}{4} \times \frac{3}{13} = \frac{45}{52}$
11. $2\frac{1}{2} \div 4\frac{1}{5} = \frac{5}{2} \times \frac{5}{21} = \frac{25}{42}$
12. $4\frac{2}{3} \div 5\frac{2}{5} = \frac{14}{3} \times \frac{5}{27} = \frac{70}{81}$
13. $3\frac{2}{3} \div 4\frac{1}{2} = \frac{11}{3} \times \frac{2}{9} = \frac{22}{27}$
14. $2\frac{1}{4} \div 4\frac{1}{5} = \frac{9}{4} \times \frac{5}{21} = \frac{15}{28}$
15. $2\frac{1}{3} \div 3\frac{2}{3} = \frac{7}{3} \times \frac{3}{11} = \frac{7}{11}$
16. $4\frac{2}{5} \div 5\frac{1}{3} = \frac{22}{5} \times \frac{3}{16} = \frac{33}{40}$
17. $4\frac{3}{5} \div 5\frac{4}{5} = \frac{23}{5} \times \frac{5}{29} = \frac{23}{29}$
18. $3\frac{3}{4} \div 6\frac{1}{2} = \frac{15}{4} \times \frac{2}{13} = \frac{15}{26}$

Divide fractions and write the answers in simplest form.

1. $7\frac{1}{2} \div 4\frac{1}{2} = \frac{15}{2} \times \frac{2}{9} = 1\frac{2}{3}$
2. $2\frac{1}{3} \div 5\frac{1}{2} = \frac{7}{3} \times \frac{2}{11} = \frac{14}{33}$
3. $5\frac{1}{2} \div 2\frac{1}{3} = \frac{11}{2} \times \frac{3}{7} = 2\frac{5}{14}$
4. $3\frac{1}{2} \div 5\frac{1}{3} = \frac{7}{2} \times \frac{3}{16} = \frac{21}{32}$
5. $6\frac{2}{3} \div 5\frac{1}{2} = \frac{20}{3} \times \frac{2}{11} = 1\frac{7}{33}$
6. $4\frac{2}{3} \div 7\frac{1}{2} = \frac{14}{3} \times \frac{3}{22} = \frac{7}{11}$
7. $5\frac{2}{3} \div 2\frac{2}{3} = \frac{17}{3} \times \frac{3}{8} = 2\frac{1}{8}$
8. $2\frac{1}{2} \div 5\frac{1}{4} = \frac{5}{2} \times \frac{4}{21} = \frac{10}{21}$
9. $5\frac{1}{5} \div 3\frac{1}{2} = \frac{26}{5} \times \frac{2}{7} = 1\frac{17}{35}$
10. $6\frac{2}{3} \div 8\frac{3}{4} = \frac{20}{3} \times \frac{4}{35} = \frac{16}{21}$
11. $8\frac{1}{4} \div 6\frac{2}{3} = \frac{33}{4} \times \frac{3}{20} = 1\frac{19}{80}$
12. $4\frac{3}{4} \div 7\frac{1}{2} = \frac{19}{4} \times \frac{2}{15} = \frac{19}{30}$
13. $7\frac{1}{3} \div 3\frac{1}{2} = \frac{22}{3} \times \frac{2}{7} = 2\frac{2}{21}$
14. $5\frac{1}{4} \div 6\frac{1}{3} = \frac{21}{4} \times \frac{3}{19} = \frac{63}{76}$
15. $9\frac{3}{5} \div 5\frac{1}{3} = \frac{48}{5} \times \frac{3}{16} = 1\frac{4}{5}$
16. $2\frac{1}{3} \div 6\frac{1}{2} = \frac{7}{3} \times \frac{2}{13} = \frac{14}{39}$
17. $8\frac{2}{5} \div 2\frac{2}{3} = \frac{42}{5} \times \frac{3}{8} = 3\frac{3}{20}$
18. $5\frac{3}{4} \div 9\frac{3}{4} = \frac{23}{4} \times \frac{4}{39} = \frac{23}{39}$

Change the following ratios to fractions and write the answer in simplest form.

Example: $2 : 3$ or 2 to $3 = \frac{2}{3}$

1. $1 : 3 = \frac{1}{3}$ 2. $2 : 7 = \frac{2}{7}$ 3. $3 : 13 = \frac{3}{13}$

4. $2 : 4 = \frac{2}{4} = \frac{1}{2}$ 5. $3 : 4 = \frac{3}{4}$ 6. $6 : 12 = \frac{6}{12} = \frac{1}{2}$

7. $1 : 5 = \frac{1}{5}$ 8. $1 : 10 = \frac{1}{10}$ 9. $4 : 13 = \frac{4}{13}$

10. $2 : 5 = \frac{2}{5}$ 11. $4 : 10 = \frac{4}{10} = \frac{2}{5}$ 12. $5 : 12 = \frac{5}{12}$

13. $3 : 7 = \frac{3}{7}$ 14. $3 : 9 = \frac{3}{9} = \frac{1}{3}$ 15. $3 : 15 = \frac{3}{15} = \frac{1}{5}$

16. $4 : 6 = \frac{4}{6} = \frac{2}{3}$ 17. $5 : 9 = \frac{5}{9}$ 18. $2 : 10 = \frac{2}{10} = \frac{1}{5}$

19. $3 : 6 = \frac{3}{6} = \frac{1}{2}$ 20. $6 : 10 = \frac{6}{10} = \frac{3}{5}$ 21. $4 : 16 = \frac{4}{16} = \frac{1}{4}$

22. $3 : 8 = \frac{3}{8}$ 23. $2 : 11 = \frac{2}{11}$ 24. $7 : 14 = \frac{7}{14} = \frac{1}{2}$

25. $3 : 9 = \frac{3}{9} = \frac{1}{3}$ 26. $4 : 12 = \frac{4}{12} = \frac{1}{3}$ 27. $8 : 24 = \frac{8}{24} = \frac{1}{3}$

Change the following ratios to fractions and write the answer in simplest form.

Example: $2 : 3$ or 2 to $3 = \frac{2}{3}$

1. $3 : 4 = \frac{3}{4}$ 2. $2 : 3 = \frac{2}{3}$ 3. $3 : 9 = \frac{3}{9} = \frac{1}{3}$

4. $2 : 6 = \frac{2}{6} = \frac{1}{3}$ 5. $4 : 5 = \frac{4}{5}$ 6. $3 : 7 = \frac{3}{7}$

7. $1 : 3 = \frac{1}{3}$ 8. $3 : 6 = \frac{3}{6} = \frac{1}{2}$ 9. $2 : 10 = \frac{2}{10} = \frac{1}{5}$

10. $3 : 5 = \frac{3}{5}$ 11. $4 : 7 = \frac{4}{7}$ 12. $4 : 6 = \frac{4}{6} = \frac{2}{3}$

13. $2 : 4 = \frac{2}{4} = \frac{1}{2}$ 14. $2 : 8 = \frac{2}{8} = \frac{1}{4}$ 15. $3 : 8 = \frac{3}{8}$

16. $2 : 7 = \frac{2}{7}$ 17. $5 : 10 = \frac{5}{10} = \frac{1}{2}$ 18. $5 : 15 = \frac{5}{15} = \frac{1}{3}$

19. $4 : 8 = \frac{4}{8} = \frac{1}{2}$ 20. $6 : 18 = \frac{6}{18} = \frac{1}{3}$ 21. $2 : 12 = \frac{2}{12} = \frac{1}{6}$

22. $5 : 9 = \frac{5}{9}$ 23. $2 : 9 = \frac{2}{9}$ 24. $6 : 10 = \frac{6}{10} = \frac{3}{5}$

25. $8 : 16 = \frac{8}{16} = \frac{1}{2}$ 26. $7 : 12 = \frac{7}{12}$ 27. $7 : 21 = \frac{7}{21} = \frac{1}{3}$

Change the following fractions to ratios.

Example: $\frac{2}{3} = 2 : 3$ or 2 to 3

1. $\frac{1}{2} = 1 : 2$ 2. $\frac{1}{3} = 1 : 3$ 3. $\frac{1}{4} = 1 : 4$

4. $\frac{2}{3} = 2 : 3$ 5. $\frac{4}{3} = 4 : 3$ 6. $\frac{5}{3} = 5 : 3$

7. $\frac{3}{2} = 3 : 2$ 8. $\frac{3}{4} = 3 : 4$ 9. $\frac{3}{5} = 3 : 5$

10. $\frac{1}{4} = 1 : 4$ 11. $\frac{5}{2} = 5 : 2$ 12. $\frac{7}{10} = 7 : 10$

13. $\frac{2}{5} = 2 : 5$ 14. $\frac{4}{7} = 4 : 7$ 15. $\frac{4}{9} = 4 : 9$

16. $\frac{5}{7} = 5 : 7$ 17. $\frac{4}{5} = 4 : 5$ 18. $\frac{10}{9} = 10 : 9$

19. $\frac{2}{9} = 2 : 9$ 20. $\frac{7}{6} = 7 : 6$ 21. $\frac{5}{4} = 5 : 4$

22. $\frac{10}{3} = 10 : 3$ 23. $\frac{4}{9} = 4 : 9$ 24. $\frac{4}{11} = 4 : 11$

25. $\frac{7}{12} = 7 : 12$ 26. $\frac{3}{10} = 3 : 10$ 27. $\frac{9}{4} = 9 : 4$

Change the following fractions to ratios.

Example: $\frac{2}{3} = 2 : 3$ or 2 to 3

1. $\frac{4}{5} = 4 : 5$ 2. $\frac{1}{5} = 1 : 5$ 3. $\frac{5}{6} = 5 : 6$

4. $\frac{2}{5} = 2 : 5$ 5. $\frac{5}{4} = 5 : 4$ 6. $\frac{7}{8} = 7 : 8$

7. $\frac{3}{4} = 3 : 4$ 8. $\frac{9}{10} = 9 : 10$ 9. $\frac{11}{12} = 11 : 12$

10. $\frac{8}{3} = 8 : 3$ 11. $\frac{11}{7} = 11 : 7$ 12. $\frac{6}{5} = 6 : 5$

13. $\frac{2}{7} = 2 : 7$ 14. $\frac{7}{2} = 7 : 2$ 15. $\frac{12}{11} = 12 : 11$

16. $\frac{5}{2} = 5 : 2$ 17. $\frac{3}{8} = 3 : 8$ 18. $\frac{8}{7} = 8 : 7$

19. $\frac{10}{9} = 10 : 9$ 20. $\frac{11}{9} = 11 : 9$ 21. $\frac{3}{14} = 3 : 14$

22. $\frac{2}{13} = 2 : 13$ 23. $\frac{15}{8} = 15 : 8$ 24. $\frac{8}{13} = 8 : 13$

25. $\frac{14}{5} = 14 : 5$ 26. $\frac{9}{16} = 9 : 16$ 27. $\frac{15}{7} = 15 : 7$

Reduce the following ratios to their lowest forms.
Example: $4:6 = 2:3$

1. $6:4 = 3:2$
2. $2:4 = 1:2$
3. $6:3 = 2:1$
4. $3:6 = 1:2$
5. $4:12 = 1:3$
6. $5:10 = 1:2$
7. $4:8 = 1:2$
8. $6:9 = 2:3$
9. $4:16 = 1:4$
10. $10:15 = 2:3$
11. $8:12 = 2:3$
12. $6:8 = 3:4$
13. $8:16 = 1:2$
14. $12:15 = 4:5$
15. $9:18 = 1:2$
16. $18:27 = 2:3$
17. $16:24 = 2:3$
18. $14:21 = 2:3$
19. $12:16 = 3:4$
20. $15:9 = 5:3$
21. $20:12 = 5:3$
22. $20:15 = 4:3$
23. $12:18 = 2:3$
24. $35:21 = 5:3$
25. $30:12 = 5:2$
26. $21:18 = 7:6$
27. $24:28 = 6:7$

Reduce the following ratios to their lowest forms.
Example: $4:6 = 2:3$

1. $12:3 = 4:1$
2. $6:10 = 3:5$
3. $12:10 = 6:5$
4. $6:18 = 1:3$
5. $42:7 = 6:1$
6. $12:32 = 3:8$
7. $21:14 = 3:2$
8. $16:48 = 1:3$
9. $40:20 = 2:1$
10. $32:8 = 4:1$
11. $24:30 = 4:5$
12. $32:12 = 8:3$
13. $18:36 = 1:2$
14. $12:8 = 3:2$
15. $9:81 = 1:9$
16. $6:8 = 3:4$
17. $18:15 = 6:5$
18. $42:56 = 3:4$
19. $3:15 = 1:5$
20. $15:35 = 3:7$
21. $13:39 = 1:3$
22. $20:8 = 5:2$
23. $18:42 = 3:7$
24. $18:4 = 9:2$
25. $15:25 = 3:5$
26. $49:28 = 7:4$
27. $22:33 = 2:3$

Reduce the following ratios to their lowest forms.
Example: $4:6 = 2:3$

1. $12:6 = 2:1$
2. $7:21 = 1:3$
3. $24:16 = 3:2$
4. $9:36 = 1:4$
5. $4:16 = 1:4$
6. $9:12 = 3:4$
7. $3:15 = 1:5$
8. $25:10 = 5:2$
9. $18:30 = 3:5$
10. $54:18 = 3:1$
11. $28:35 = 4:5$
12. $42:12 = 7:2$
13. $20:24 = 5:6$
14. $12:8 = 3:2$
15. $45:63 = 5:7$
16. $21:49 = 3:7$
17. $5:45 = 1:9$
18. $35:65 = 7:13$
19. $40:15 = 8:3$
20. $32:56 = 4:7$
21. $24:36 = 2:3$
22. $64:48 = 4:3$
23. $48:32 = 3:2$
24. $52:39 = 4:3$
25. $6:27 = 2:9$
26. $72:63 = 8:7$
27. $33:55 = 3:5$

Find equivalent ratios.
Example: $1:2 = 2:4 = 3:6 = 4:8$

1. $2:3 = 4:\underline{6} = 6:\underline{9} = 8:\underline{12} = 12:\underline{18} = 24:\underline{36}$
2. $1:3 = \underline{2}:6 = \underline{3}:9 = \underline{5}:15 = \underline{6}:18 = \underline{9}:27$
3. $3:2 = 9:\underline{6} = 15:\underline{10} = \underline{18}:12 = \underline{24}:16 = 30:\underline{20}$
4. $1:4 = 2:\underline{8} = \underline{3}:12 = 5:\underline{20} = \underline{6}:24 = \underline{8}:32$
5. $4:3 = 12:\underline{9} = 16:\underline{12} = \underline{24}:18 = \underline{28}:21 = 36:\underline{27}$
6. $2:5 = \underline{8}:20 = \underline{10}:25 = 14:\underline{35} = 16:\underline{40} = 22:\underline{55}$
7. $3:4 = 9:\underline{12} = 15:\underline{20} = \underline{18}:24 = \underline{24}:32 = 36:\underline{48}$
8. $1:5 = \underline{2}:10 = \underline{3}:15 = \underline{5}:25 = \underline{6}:30 = 8:\underline{40}$
9. $5:2 = 10:\underline{4} = \underline{20}:8 = 25:\underline{10} = \underline{30}:12 = 45:\underline{18}$
10. $2:7 = 4:\underline{14} = 8:\underline{28} = \underline{10}:35 = \underline{14}:49 = 16:\underline{56}$
11. $8:3 = 24:\underline{9} = 32:\underline{12} = \underline{48}:18 = \underline{56}:21 = 72:\underline{27}$
12. $3:5 = 6:\underline{10} = 9:\underline{15} = 12:\underline{20} = \underline{18}:30 = \underline{27}:45$
13. $5:6 = \underline{15}:18 = 20:\underline{24} = 25:\underline{30} = \underline{35}:42 = \underline{40}:48$

Find equivalent ratios.
Example:　$1:2 = 2:4 = 3:6 = 4:8$

1. $2:3 = 4:\underline{6} = 6:\underline{9} = 8:\underline{12} = 12:\underline{18} = 16:\underline{24}$

2. $2:5 = \underline{6}:15 = 10:\underline{25} = \underline{14}:35 = 16:\underline{40} = 26:\underline{65}$

3. $4:1 = 8:\underline{2} = \underline{16}:4 = 20:\underline{5} = \underline{28}:7 = 36:\underline{9}$

4. $3:2 = \underline{9}:6 = 12:\underline{8} = \underline{15}:10 = 21:\underline{14} = \underline{27}:18$

5. $10:4 = 5:\underline{2} = 15:\underline{6} = \underline{40}:16 = 30:\underline{12} = \underline{80}:32$

6. $3:7 = 9:\underline{21} = \underline{12}:28 = \underline{15}:35 = 18:\underline{42} = \underline{33}:77$

7. $5:4 = \underline{15}:12 = \underline{20}:16 = \underline{25}:20 = 30:\underline{24} = 45:\underline{36}$

8. $4:3 = 8:\underline{6} = 12:\underline{9} = \underline{16}:12 = \underline{24}:18 = 28:\underline{21}$

9. $4:6 = 2:\underline{3} = 6:\underline{9} = \underline{8}:12 = 12:\underline{18} = \underline{16}:24$

10. $7:4 = 28:\underline{16} = \underline{35}:20 = 42:\underline{24} = 56:\underline{32} = \underline{63}:36$

11. $8:4 = \underline{4}:2 = 2:\underline{1} = 10:\underline{5} = \underline{12}:6 = 16:\underline{8}$

12. $3:1 = 9:\underline{3} = 12:\underline{4} = \underline{15}:5 = \underline{24}:8 = 33:\underline{11}$

13. $12:2 = 6:\underline{1} = 18:\underline{3} = \underline{30}:5 = \underline{72}:12 = \underline{96}:16$

Find equivalent ratios.
Example:　$1:2 = 2:4 = 3:6 = 4:8$

1. $2:1 = \underline{4}:2 = 10:\underline{5} = 12:\underline{6} = 18:\underline{9} = \underline{30}:15$

2. $3:2 = \underline{6}:4 = \underline{9}:6 = 15:\underline{10} = 27:\underline{18} = \underline{39}:26$

3. $10:4 = 5:2 = \underline{25}:10 = 35:\underline{14} = \underline{45}:18 = 70:\underline{28}$

4. $4:6 = \underline{2}:3 = 10:\underline{15} = 12:\underline{18} = \underline{16}:24 = \underline{30}:45$

5. $6:10 = 3:\underline{5} = 12:\underline{20} = \underline{21}:35 = 24:\underline{40} = \underline{30}:50$

6. $12:4 = \underline{3}:1 = 15:\underline{5} = 24:\underline{8} = \underline{36}:12 = 45:\underline{15}$

7. $15:12 = 5:\underline{4} = \underline{10}:8 = 30:\underline{24} = 45:\underline{36} = \underline{60}:48$

8. $9:21 = \underline{3}:7 = \underline{12}:28 = \underline{24}:56 = \underline{27}:63 = \underline{36}:84$

9. $16:12 = 4:\underline{3} = 8:\underline{6} = 28:\underline{21} = 40:\underline{30} = 56:\underline{42}$

10. $10:45 = \underline{2}:9 = 12:\underline{54} = 14:\underline{63} = \underline{18}:81 = \underline{22}:99$

11. $21:24 = 7:\underline{8} = 14:\underline{16} = \underline{35}:40 = \underline{56}:64 = 84:\underline{96}$

12. $36:45 = 4:\underline{5} = 12:\underline{15} = \underline{20}:25 = 28:\underline{35} = 60:\underline{75}$

13. $14:12 = \underline{7}:6 = 21:\underline{18} = \underline{42}:36 = 49:\underline{42} = 63:\underline{54}$

Evaluate the following exponents.
Example:　$5^2 = 5 \times 5 = 25,$　　$4^3 = 4 \times 4 \times 4 = 64,$　　$2^4 = 2 \times 2 \times 2 \times 2 = 16$

1. $2^2 = 4$　　　　2. $2^3 = 8$　　　　3. $4^2 = 16$

4. $4^2 = 16$　　　　5. $3^3 = 27$　　　　6. $7^2 = 49$

7. $3^2 = 9$　　　　8. $4^2 = 16$　　　　9. $3^4 = 81$

10. $8^2 = 64$　　　11. $2^5 = 32$　　　12. $6^2 = 36$

13. $2^6 = 64$　　　14. $6^2 = 36$　　　15. $3^5 = 243$

16. $7^2 = 49$　　　17. $10^2 = 100$　　18. $6^3 = 216$

19. $2^7 = 128$　　　20. $5^3 = 125$　　　21. $2^6 = 64$

22. $9^2 = 81$　　　23. $4^4 = 256$　　　24. $2^8 = 256$

25. $11^2 = 121$　　26. $7^3 = 343$　　　27. $12^2 = 144$

Evaluate the following exponents.
Example:　$5^2 = 5 \times 5 = 25,$　　$4^3 = 4 \times 4 \times 4 = 64,$　　$2^4 = 2 \times 2 \times 2 \times 2 = 16$

1. $5^2 = 25$　　　　2. $3^2 = 9$　　　　3. $4^2 = 16$

4. $3^3 = 27$　　　　5. $7^2 = 49$　　　　6. $2^6 = 64$

7. $8^2 = 64$　　　　8. $9^2 = 81$　　　　9. $5^3 = 125$

10. $6^3 = 216$　　　11. $2^7 = 128$　　　12. $2^9 = 512$

13. $2^{10} = 1024$　　14. $4^3 = 64$　　　15. $8^3 = 512$

16. $3^5 = 243$　　　17. $6^3 = 216$　　　18. $12^2 = 144$

19. $4^4 = 256$　　　20. $11^2 = 121$　　21. $15^2 = 225$

22. $17^2 = 289$　　23. $14^2 = 196$　　24. $16^2 = 256$

25. $18^2 = 324$　　26. $19^2 = 361$　　27. $20^2 = 400$

Evaluate the following exponents.

Example: $5^2 = 5 \times 5 = 25$, $4^3 = 4 \times 4 \times 4 = 64$, $2^4 = 2 \times 2 \times 2 \times 2 = 16$

1. $5^2 = 25$	2. $7^2 = 49$	3. $4^2 = 16$
4. $3^3 = 27$	5. $4^3 = 64$	6. $5^3 = 125$
7. $2^5 = 32$	8. $11^2 = 121$	9. $9^2 = 81$
10. $2^9 = 512$	11. $6^3 = 216$	12. $12^2 = 144$
13. $6^2 = 36$	14. $2^7 = 128$	15. $10^2 = 100$
16. $14^2 = 196$	17. $8^2 = 64$	18. $7^3 = 343$
19. $15^2 = 225$	20. $13^2 = 169$	21. $2^8 = 256$
22. $21^2 = 441$	23. $16^2 = 256$	24. $10^3 = 1000$
25. $9^3 = 729$	26. $25^2 = 625$	27. $22^2 = 484$

Evaluate the following exponents.

Example: $5^2 = 5 \times 5 = 25$, $4^3 = 4 \times 4 \times 4 = 64$, $2^4 = 2 \times 2 \times 2 \times 2 = 16$

1. $2^6 = 64$	2. $4^3 = 64$	3. $2^5 = 32$
4. $6^3 = 216$	5. $3^5 = 243$	6. $8^3 = 512$
7. $13^2 = 169$	8. $2^{10} = 1024$	9. $10^2 = 100$
10. $14^2 = 196$	11. $15^2 = 225$	12. $16^2 = 256$
13. $2^9 = 512$	14. $17^2 = 289$	15. $24^2 = 576$
16. $20^2 = 400$	17. $3^4 = 81$	18. $7^2 = 49$
19. $30^2 = 900$	20. $10^3 = 1000$	21. $18^2 = 324$
22. $9^3 = 729$	23. $25^2 = 625$	24. $23^2 = 529$
25. $12^2 = 144$	26. $5^3 = 125$	27. $17^2 = 289$

Evaluate the following exponents.

Example: $2^4 + 3^2 = 16 + 9 = 25$

1. $2^3 + 3^2 = 17$	2. $2^2 + 5^2 = 29$	3. $5^2 + 4^2 = 41$
4. $2^4 + 3^3 = 43$	5. $4^2 + 6^2 = 52$	6. $7^2 + 4^3 = 113$
7. $2^3 + 4^2 = 24$	8. $3^3 + 6^2 = 63$	9. $10^2 + 9^2 = 181$
10. $5^2 + 3^3 = 52$	11. $4^3 + 2^3 = 72$	12. $3^5 + 4^3 = 307$
13. $2^6 + 3^4 = 145$	14. $2^5 + 3^4 = 113$	15. $12^2 + 5^2 = 169$
16. $6^2 + 7^2 = 85$	17. $3^4 + 5^3 = 206$	18. $2^7 + 2^6 = 192$
19. $9^2 + 8^2 = 145$	20. $12^2 + 2^3 = 152$	21. $10^2 + 13^2 = 269$
22. $11^2 + 5^2 = 146$	23. $8^2 + 12^2 = 208$	24. $3^3 + 3^4 = 108$
25. $2^8 + 2^9 = 768$	26. $5^3 + 6^3 = 341$	27. $10^2 + 10^3 = 1100$

Evaluate the following exponents.

Example: $2^4 + 3^2 = 16 + 9 = 25$

1. $3^4 + 4^2 = 97$	2. $8^2 + 9^2 = 145$	3. $3^2 + 4^2 = 25$
4. $12^2 + 5^2 = 169$	5. $5^2 + 10^2 = 125$	6. $9^2 + 11^2 = 202$
7. $8^2 + 4^3 = 128$	8. $2^4 + 2^5 = 48$	9. $6^2 + 8^2 = 100$
10. $3^3 + 5^3 = 152$	11. $12^2 + 14^2 = 340$	12. $2^8 + 14^2 = 452$
13. $2^6 + 9^2 = 145$	14. $5^3 + 10^2 = 225$	15. $15^2 + 20^2 = 625$
16. $8^2 + 15^2 = 289$	17. $6^3 + 3^5 = 459$	18. $6^2 + 7^2 = 85$
19. $13^2 + 19^2 = 530$	20. $11^2 + 12^2 = 265$	21. $10^2 + 20^2 = 500$
22. $2^5 + 2^7 = 160$	23. $2^9 + 2^8 = 768$	24. $12^2 + 17^2 = 433$
25. $7^3 + 16^2 = 599$	26. $5^3 + 9^3 = 854$	27. $8^3 + 16^2 = 768$

Evaluate the following exponents.
Example: $3^4 - 4^3 = 81 - 64 = 17$

1. $6^2 - 4^2 = 20$ 2. $9^2 - 8^2 = 17$ 3. $8^2 - 4^2 = 48$

4. $4^3 - 2^4 = 48$ 5. $11^2 - 3^4 = 40$ 6. $3^4 - 7^2 = 32$

7. $10^2 - 8^2 = 36$ 8. $6^3 - 4^3 = 152$ 9. $7^3 - 2^6 = 279$

10. $5^3 - 5^2 = 100$ 11. $14^2 - 13^2 = 27$ 12. $21^2 - 12^2 = 297$

13. $7^2 - 3^2 = 40$ 14. $2^8 - 5^3 = 131$ 15. $10^2 - 6^2 = 64$

16. $2^5 - 4^2 = 16$ 17. $15^2 - 13^2 = 56$ 18. $2^9 - 6^3 = 296$

19. $12^2 - 6^2 = 108$ 20. $3^5 - 15^2 = 18$ 21. $19^2 - 2^6 = 297$

22. $6^3 - 9^2 = 135$ 23. $14^2 - 3^4 = 115$ 24. $16^2 - 14^2 = 60$

25. $2^7 - 4^3 = 64$ 26. $20^2 - 17^2 = 111$ 27. $4^4 - 5^3 = 131$

Evaluate the following exponents.
Example: $3^4 - 4^3 = 81 - 64 = 17$

1. $9^2 - 3^2 = 72$ 2. $10^2 - 4^2 = 84$ 3. $7^2 - 3^2 = 40$

4. $3^4 - 6^2 = 45$ 5. $4^3 - 2^5 = 39$ 6. $12^2 - 9^2 = 63$

7. $5^3 - 2^6 = 61$ 8. $12^2 - 7^2 = 95$ 9. $3^4 - 4^2 = 65$

10. $10^2 - 5^2 = 75$ 11. $7^3 - 15^2 = 118$ 12. $15^2 - 11^2 = 104$

13. $2^8 - 4^3 = 192$ 14. $2^8 - 4^3 = 192$ 15. $2^8 - 5^3 = 131$

16. $5^3 - 4^3 = 61$ 17. $10^3 - 2^9 = 488$ 18. $7^3 - 2^7 = 215$

19. $11^2 - 9^2 = 40$ 20. $8^3 - 6^3 = 296$ 21. $3^5 - 5^3 = 118$

22. $6^3 - 13^2 = 47$ 23. $4^4 - 3^5 = 13$ 24. $4^4 - 15^2 = 31$

25. $15^2 - 2^7 = 97$ 26. $6^3 - 5^3 = 91$ 27. $9^3 - 25^2 = 104$

Evaluate the following exponents.
Example: $2^3 \times 3^2 = 8 \times 9 = 72$

1. $2^2 \times 3^2 = 36$ 2. $3^2 \times 3^3 = 243$ 3. $7^2 \times 2^2 = 196$

4. $3^2 \times 2^3 = 72$ 5. $4^2 \times 4^2 = 256$ 6. $8^2 \times 2^3 = 512$

7. $2^3 \times 4^2 = 128$ 8. $4^3 \times 2^2 = 256$ 9. $2^4 \times 3^3 = 432$

10. $3^2 \times 4^2 = 144$ 11. $2^5 \times 2^3 = 256$ 12. $10^2 \times 2^2 = 400$

13. $2^4 \times 2^2 = 64$ 14. $5^2 \times 2^2 = 100$ 15. $7^2 \times 4^2 = 784$

16. $2^4 \times 3^2 = 144$ 17. $6^2 \times 3^2 = 324$ 18. $4^3 \times 2^3 = 512$

19. $2^5 \times 2^2 = 128$ 20. $5^2 \times 2^3 = 200$ 21. $2^5 \times 3^3 = 864$

22. $2^5 \times 3^2 = 288$ 23. $3^3 \times 2^3 = 216$ 24. $4^2 \times 3^3 = 432$

25. $2^6 \times 2^2 = 256$ 26. $5^3 \times 2^3 = 1000$ 27. $2^5 \times 2^5 = 1024$

Evaluate the following exponents.
Example: $2^3 \times 3^2 = 8 \times 9 = 72$

1. $2^3 \times 4^2 = 128$ 2. $2^4 \times 2^4 = 256$ 3. $5^2 \times 2^6 = 1600$

4. $3^3 \times 2^3 = 216$ 5. $10^2 \times 2^3 = 800$ 6. $2^5 \times 2^4 = 512$

7. $2^4 \times 4^2 = 256$ 8. $11^2 \times 2^2 = 484$ 9. $3^2 \times 2^5 = 288$

10. $5^2 \times 2^2 = 100$ 11. $4^3 \times 2^4 = 1024$ 12. $4^2 \times 2^4 = 256$

13. $6^2 \times 2^3 = 288$ 14. $2^5 \times 2^3 = 256$ 15. $2^4 \times 5^3 = 2000$

16. $3^3 \times 3^3 = 729$ 17. $3^4 \times 2^3 = 648$ 18. $2^3 \times 4^4 = 2048$

19. $4^3 \times 2^2 = 256$ 20. $2^6 \times 2^4 = 1024$ 21. $13^2 \times 2^2 = 676$

22. $5^3 \times 2^3 = 1000$ 23. $3^5 \times 2^2 = 972$ 24. $3^5 \times 2^3 = 1944$

25. $6^3 \times 2^2 = 864$ 26. $6^3 \times 2^2 = 864$ 27. $7^3 \times 2^2 = 1372$